SpringerBriefs in Operations Management

Series Editor

Suresh P. Sethi
The University of Texas at Dallas, TX, USA

More information about this series at http://www.springer.com/series/13082

Petri Helo • Angappa Gunasekaran
Anna Rymaszewska

Designing and Managing Industrial Product-Service Systems

 Springer

Petri Helo
Networked Value Systems, Department of
  Production
University of Vaasa
Vaasa, Finland

Angappa Gunasekaran
Charlton College of Business
University of Massachusetts, Dartmouth
North Dartmouth, MA, USA

Anna Rymaszewska
Networked Value Systems, Department of
  Production
University of Vaasa
Vaasa, Finland

ISSN 2365-8320　　　　　　　ISSN 2365-8339　(electronic)
SpringerBriefs in Operations Management
ISBN 978-3-319-40429-5　　　ISBN 978-3-319-40430-1　(eBook)
DOI 10.1007/978-3-319-40430-1

Library of Congress Control Number: 2016945169

© Springer International Publishing Switzerland 2017
This work is subject to copyright. All rights are reserved by the Publisher, whether the whole or part of the material is concerned, specifically the rights of translation, reprinting, reuse of illustrations, recitation, broadcasting, reproduction on microfilms or in any other physical way, and transmission or information storage and retrieval, electronic adaptation, computer software, or by similar or dissimilar methodology now known or hereafter developed.
The use of general descriptive names, registered names, trademarks, service marks, etc. in this publication does not imply, even in the absence of a specific statement, that such names are exempt from the relevant protective laws and regulations and therefore free for general use.
The publisher, the authors and the editors are safe to assume that the advice and information in this book are believed to be true and accurate at the date of publication. Neither the publisher nor the authors or the editors give a warranty, express or implied, with respect to the material contained herein or for any errors or omissions that may have been made.

This Springer imprint is published by Springer Nature
The registered company is Springer International Publishing AG Switzerland

Contents

1	**Introduction**	1
	1.1 Justification and Importance	3
	1.1 Challenge of Transition	4
2	**Servitization: Service Infusion in Manufacturing**	5
	2.1 Servitization: Transition Toward Services	5
	2.2 Definition of the Concept	8
	2.3 Commodization and Decommodization	9
	2.4 Manufacturing and Service Organizations	10
	2.5 New Service Development	11
	2.5.1 Service Blueprinting	11
	2.5.2 Service Innovation and Business Models	13
	2.6 Product Life Cycle as Platform of Servitization	14
	2.6.1 Life Cycle for Machinery Delivery	15
	2.6.2 Software Process Life Cycle	16
	2.7 The Servitization Paradox	17
3	**Integrated Product-Service Systems**	19
	3.1 Types of Industrial Services	19
	3.2 Product-Service System	20
	3.3 Benefits of PSS	21
	3.4 Characteristics of PSS	22
	3.5 Barriers to PSS	23
	3.6 Steps Toward Integrated Product-Service Systems	24
4	**Strategic Improvements Through Industrial Services**	27
	4.1 Classification of Industrial Services	28
	4.2 From Goods-Dominant to Service-Dominant Logic	33
	4.3 Goods-Dominant Logic vs. Service-Dominant Logic	34

5	**Improving Marketing and Operations Strategy Through Industrial Services**	**37**
	5.1 Coproduction of Service Offering and Customer Experience Management	37
	5.2 Increasing Competitiveness Through Service-Dominant Logic	38
	5.3 Global Service Strategies: Service as a Means of Expanding Business	38
	5.4 Service Supply Chain Structure	41
6	**Service Delivery**	**43**
	6.1 Service Delivery Concept	43
	6.2 Service Delivery System Design	43
	6.3 Customers' Roles in Service Delivery	44
	6.4 Customer Expectation of Industrial Services	45
7	**Managing Service Delivery**	**49**
	7.1 Service-Level Agreements	49
	7.2 Performance Measurement	51
	7.2.1 Service Delivery Quality	52
	7.3 Installed Base Management	53
	7.4 Enterprise Asset Management	55
8	**Role of Technology in Servitization**	**57**
	8.1 Technology and Servitization	57
	8.2 Internet and Connected Products	58
	8.2.1 Architecture of Smart Physical Products	59
	8.2.2 Remote Management Systems in Service Products	60
	8.3 Industrial Standards for Managing Service Platforms	62
	8.4 Condition-Based Maintenance	64
	8.5 Cloud-Based Services and Portals	68
	8.6 Big Data Analytics	69
	8.7 Applification	70
9	**Pricing Decisions: From Ownership to Subscription**	**73**
	9.1 Subscription Services	73
	9.2 CAPEX and OPEX	75
	9.3 The Subscription Economy	77
	9.4 Pricing Models	77
	9.5 Freemium Pricing Model	79
	9.6 Software Ecosystem Model	79
	9.6.1 Directed Approach	80
	9.6.2 Undirected Approach	81
	9.6.3 Tiers of Developers	81
10	**Value Chain Effects**	**83**
	10.1 Cocreation and Coproduction of Value	83
	10.2 Supply Chains and Networks	84

	10.3 Vertical and Horizontal Integration	84
	10.4 Moving Downstream in a Value Chain	86
11	**Conclusions**	89

Acknowledgment 91

References 93

Index 99

Chapter 1
Introduction

The service supply chain has become an integral and essential part of supply chain design. The service element of business is becoming increasingly important for companies operating in manufacturing. This trend has led companies to strive toward a greater focus on service aspect of the product offering (Fang et al. 2008).

Companies designing and manufacturing machinery are increasingly introducing new remote monitoring systems to enhance the product functionality, from a physical product to a product containing information and service features. By doing so they are connecting to customer asset and coming closer to customer operations. Such an approach allows for connecting to customer's assets and shortening the distance between operations of a manufacturing company and its customer. In order to be able to achieve that, new capabilities are required. These capabilities are not limited to manufacturing technology but understanding how to run customer's operations effectively. The focus of delivery is shifting from tangibles toward intangibles as product-service systems are offered as integrated packaged product bundles. This development is led by Western companies who face fierce competition from low-cost countries. Since competing on cost might not necessarily be a feasible nor successful strategy, more advanced, technology-based approaches are needed.

Due to the aforementioned factors, the global industry structure is transitioning from the "industrial economy" to something which may be referred to as a "service economy (Heinzel et al. 2007; Tuli et al. 2007)." The service supply chain (SSC) has become a more frequently researched topic (Lin et al. 2010). The design and coordination of the supply chain management, including distribution network, are critical for any organization striving to deliver their products and services productively. Supply chain management of today goes beyond purchasing, distribution, and logistics to encompass the new areas such as supplier relationships, supplier network structure, and collaboration with suppliers.

Already in year 2004, private service sector accounted for close to 70 % of the GDP (Strassner and Howells 2005). This indicates the importance of service economy and its industries.

This report is focused on analyzing how transition from products to services can be managed and how supply chains could be adjusted to the new norm. In the consecutive chapters, product-service system structures are analyzed, and servitization—the service infusion process—is discussed, followed by the presentation of industrial services as marketing and operations strategy. Next, Chap. 6 discusses how the actual operations take place. Chapter 8 aims at exploring how connected assets are utilized by product vendors in value creation. Then, transition from ownership to subscriptions is analyzed in pricing decision section. Chapter 10 is focused on providing an overview of the mechanisms through which industrial companies are shortening the distance to end users and aim for better position in value chain. The report closes with conclusions where also theoretical and empirical implications are provided.

With industrial services supply chain management being so prominent, it is important to build an understanding of its issues and complexities. Service operations management encompasses managing of processes that deliver services crucial to achieving continuous improvement that reaches beyond simply delivering services to customers. Traditional notion of supply chains focuses on products and shorter-term profit with the objective of achieving competitive lean or agile supply. From managerial perspective, the performance of such product-centric supply chain is measured in terms of as cost, quality, productivity, flexibility, dependability, and responsiveness. Nowadays, the focus is shifting from product-related services to those that offer a solution to customer problems or requirements, by considering the entire product life cycle.

According to Edgett (1994) developing industrial customer services should be controlled and managed as new product development in the goods-dominant industries. Markets should be divided into segments and service products should be developed to fit in with the needs and requirements for each segment. Sharma and Lambert (1994) have suggested industrial market segmentation and use of customer group specific market approached. Detailed technical specifications including required competences and business cases are then designed according to analyzed and prioritized needs. Achieving generic service offerings aiming to fulfill the needs for all sectors is challenging. Unique service packages may be based on configurable product offerings which consist of selections, parameterization, and optional elements. The purpose of classification process is to create service products that lead to higher level of customer satisfaction (Burger and Cann 1995).

Maintaining service level for existing customers has gained importance as relationship marketing has become a standard approach in industrial marketing management (Morgan and Hunt 1994). On the other side, manufacturing companies have developed long-term partnerships with the supply and distribution side. Defined and managed service offering is then operationalized in the delivery process. Supply chain management combines delivery and supply chains in the analysis. This perspective focuses on delivery and aims to improve quality, reliability, cost, and time-related performance indicators.

1.1 Justification and Importance

Three major domains can be identified in the frontiers of research in the field of supply chain management in industrial services. Firstly, the business environment has shown the importance of service offerings and what type of products are being considered. Secondly, standardization coming from IT development is driving how technical infrastructures should be built for managing services. Thirdly, products offered by leading industrial companies show innovative approaches to business delivery.

Business importance of industrial services has been widely recognized by mature companies operating in machine-building industry. Companies seek growth not only by selling service products to existing customers but also by offering tangible products to new customers. Table 1.1 presents a study performed by Deloitte Research, proving how top 90 % companies in global businesses gather more than 50 % of their revenues from service and spare parts. For example, industrial average in the aerospace and defense is 47 % and automotive and commercial vehicles sector read 37 % of the total revenue. Surprisingly, high technology and telecommunication industry gains only 19 % of the revenue from services. The differences are most likely caused by criticality of maintenance and focus on asset life cycle but also on industry-specific traditions and culture. Common for all these industry categories is that the top 90 % fractile of the companies have more than 50 % of the revenue originating from the services. In conclusion, leading companies at all fields are able to reap the benefits from service business potential.

Several European machine-building companies are at the same ranges, but many smaller companies are also taking first steps toward this direction. From supply chain management point of view, the strategies and tactics need to be tailored to match those approaches which cannot be solely reliant on global presence.

Table 1.1 Revenue shares of selected global industries

	Share of service and spare part business in overall sales	
Global industry	Average (%)	Top 90 %
Aerospace and defense	47	>50 %
Automotive and commercial vehicles	37	>50 %
Diversified manufacturing and industrial products	20	>50 %
High technology and telecommunication equipment	19	>50 %
Life sciences/medical devices	21	>50 %
All companies	26	>50 %

Source: Deloitte Research—Global Service and Parts Management Benchmark Survey

1.2 Challenge of Transition

Companies beyond the automotive and aerospace are looking for opportunities to increase service offering. The transition takes place in several other industries too. Many industrial companies have perceived the fact that surviving in the global competition requires industrial companies to shift from physical products to a set of value-added services. The companies thus need to position themselves as service vendors or solution providers (Vargo and Lusch 2008). Large multinational companies are ahead of this transition and medium-sized companies are following. The motivations are nevertheless similar. Higher customer demand and increased revenue stream (Oliva and Kallenberg 2003), stability of revenues (Mathe and Shapiro 1993), and overall competition pressure from the market (Gaubauwer and Friedli 2005) are typical driving factors.

Progress of advances of technology is an important driver. Pervasive communication capacities of smart devices are building blocks for service development. In the following chapters, we deal with the transition mechanisms. Firstly, service infusion in manufacturing is discussed, followed by integrated product-service system definitions as well as industrial services as marketing and operations strategy. Then strategic improvements, service delivery, and management of service are presented. Role of technologies and pricing decisions as part of transition from ownership to subscription are presented. Finally, value chain effects from the transition are discussed along with the concluding section.

Chapter 2
Servitization: Service Infusion in Manufacturing

Extending a company's offer beyond manufacturing has become an effective way of increasing profits and staying ahead of competitors. It is difficult to pinpoint the emergence of servitization. Many researchers in the field, such as Vandermerwe and Rada (1989), claim that servitization is occurring on a global scale, and manufacturing organizations are unable to avoid the transition toward extending their offering.

2.1 Servitization: Transition Toward Services

The idea of servitization is easier to grasp when service is defined in the first place. Grönroos (2008, p. 52) defines service as follows: "a process consisting of a series of more or less intangible activities that normally, but not necessarily always, take place in interactions between the customer and service employees and/or physical resources or goods and/or systems of the service provider, which are provided as solution to customer problems." This widely accepted definition is quite broad and generally focuses on interaction perspective. Servitization is about transition from the physical world to intangibles. For this reason, it must be distinguished from products. According to Van Looy et al. (2003), the features of services include the following items:

- Customer participation, which refers to interaction between supply and demand, value co-production.
- Simultaneity of supply and consumption occurring.
- Perishability of the operation once process is complete and stock lacking.
- Intangibility—nonphysical nature of the service.
- Heterogeneity—customer participation may induce variability and reduce control.
- Nontransferable ownership.

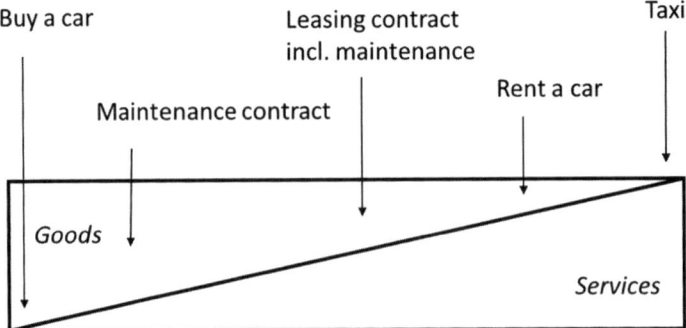

Fig. 2.1 Products are a "bundle" of goods and services. Servitization refers to the transformation toward the *right side* of the picture

According to Neely et al. (2011), there can be different types of services that are offered as the extension to a tangible product, such as:

- System services combining tailored product offerings into customer solutions.
- Design and development services supporting the operational performance of the product utilization.
- Maintenance and support to extend product life cycle.
- Retail and distribution services to improve the accessibility.
- Commissioning and installation services to complete implementation and operationalization.

The spectrum from pure physical products, the goods, to pure intangible services is quite wide ranging. In many cases both elements are involved to a certain extent. Servitization is about moving in this spectrum toward the service end (Edvardsson et al. 2005; Gebauer 2008; Gebauer and Friedli 2005). Ren and Gregory (2007) defined servitization as "a change process wherein manufacturing companies embrace service orientation and/or develop more and better services, with the aim to satisfy customer's needs, achieve competitive advantage and enhance firm performance." Another practical definition is given by Ward and Graves (2005): "increasing the range of services offered by a manufacturer."

Figure 2.1 illustrates the spectrum from goods to services and examples of transportation-related solutions for markets. On the goods side, buying a car could be an example of an investment. Maintenance contracts or warranty services could add an intangible part. A leasing contract would add the finance service element and reduce the ownership aspect. Car rental would present short-term commitment and bring the focus onto actual operations. A taxi cab could be presented as a pure service example in this case. The spectrum is not discrete by nature but smoothly continuous. Due to the dual nature and coexistence of the elements, a product-service system (PSS) or product-service bundle concept is used.

Table 2.1 shows the differences between physical products and intangible services from different perspectives. The focus is very different in terms of delivery. Life cycle management of assets brings the operational cost aspects to the forefront

2.1 Servitization: Transition Toward Services

Table 2.1 Products and services—characteristics

	Products	Services
Scope of delivery	Physical delivery	Physical delivery
		Maintenance
		Support
Scope of life cycle	Delivery	Delivery
		Operations
		Reverse logistics
Financial transactions	On delivery	Per use
Number of supply chain transactions	Few	Frequently
Cost accounting focus	Capital cost of investment	Capital cost
		Operational costs
		Expected life cycle
		Interest rates
Number of supply chain end points	Few delivery addresses	Large number of actual sites of use
End point types	Customer delivery to next level in supply chain	End users
Demand pattern	Fluctuations based on investments	Stable based on actual use
Supply chain efficiency focus	Delivery	Response to end-customer requests
Unit of analysis	Customers	Installed base

Source: Neely et al. (2011)

compared to investment costs and capital expenses. Managing installed base and actual operation transactions instead of delivering the physical machinery is another transition that can be perceived.

Many manufacturing companies have realized that the gradual enriching of their offering is at least a promising, if not a necessary, development path. The shift toward servitization is in itself nothing unique or recent. However, the scale and speed of the servitization has been increasing considerably. Although servitization is generally referred to as a Western phenomenon (Baines 2013), many less industrialized countries are following. According to Neely et al. (2011), even China, a country traditionally dependent on manufacturing as the main driver of economic growth, has now noticed a sharp increase in GDP generated from services.

Neely et al. (2011) also look at servitization in different countries by examining the percentage of manufacturing companies that have shifted to offering services between the years 2007 and 2011. Figure 2.2 presents their findings and it shows that the degree of servitization has increased in most of the featured countries. In the data the USA and Finland are the leading countries followed by European countries and Singapore and Malaysia in Asia.

Transformation is emerging in consumer business and the business to business side is adapting the models. Some interesting benchmark in business models is

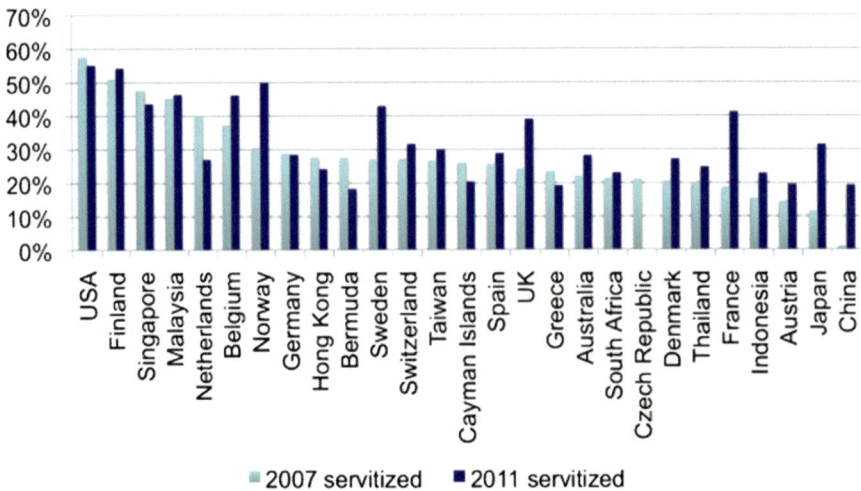

Fig. 2.2 Percentage of manufacturing companies that have shifted to offering services in the years 2007–2011. Adapted from Neely et al. (2011)

proposed by Botsman (2010), who suggests a transition from how a car is seen as an owned product toward something bought as a service, as illustrated by the examples of Zipcar—a company acquired by the Avis Budget Group. The ownership of an asset-related servitization phenomenon itself would be similar to how the music industry has seen a transition from physical products (CDs) to intangible electronic files (iTunes) and ultimately subscription service (Spotify). There is a gap between what technologies enable and what companies are actually doing in the field of organization and managing supply chains in industrial services.

What industry is about to see is that transformation is also taking place in the business to business environment. Consumer businesses are teaching the masses the benefits of technology-enabled services. The process of transformation is enabled by information and communication technologies and real-time access to data. Supply chain structures and key performance indicators are affected by these changes. For these reasons industrial service models are shaping future supply chain architectures.

2.2 Definition of the Concept

Vandermerwe and Rada (1989) provided the first definition of servitization and sparked a stream of research in the field. The authors perceive servitization as "market packages or bundles of customer-focused combinations of goods, services, support, self-service and knowledge." Baines et al. (2009) define servitization as "adding value by adding services to products." According to Baines et al. (2009), the main drive toward servitization is that the integrated product-service offering is an excellent means of building a long-term competitive advantage. The authors also provide a concise overview of the most significant definitions of servitization over the period between the years 1998 and 2007. While the core meaning has remained unchanged, the idea has been described in terms of a "trend," "integrated bundle," "strategy," and "change process." Therefore, it can be stated that the definitions of servitization differ in their scope. Baines et al. (2009) conclude that servitization is "the innovation of an organizations capabilities and processes to better create mutual value through a shift from selling products to selling product-service systems (PSS)."

Baines et al. (2009) define product-service system as "an integrated combination of products and services that deliver value in use." Although those two streams of research have been developing separately, Baines et al. (2009) claim that nowadays PSS can be considered as a subset of servitization research. Baines et al. (2009) propose a summary of the most important definitions of servitization (Fig. 2.2). Baines et al. (2009) examine the main features of servitization. The authors outline the following:

- Strong customer centricity
- A multi-vendor approach to delivery of the solutions tailored to individual customers
- A shift from product-oriented services toward user process-oriented services
- A transition in the nature of customer's interactions—from transaction based to relationship based

There are various factors motivating the decision on servitization. The three main goals as identified by Baines et al. (2009) include financial, strategic (creating the competitive edge), and marketing. According to the authors, financial drivers translate into higher profit margins and stability of income, while strategic drivers are all about creating a competitive advantage. According to Baines et al. (2009), competitive advantage can be gained by using services to differentiate manufacturing offerings as well as provide crucial competitive opportunities.

2.3 Commodization and Decommodization

Commodization and decommodization are concepts related to the servitization process. According to Matthyssens and Vandenbempt (2008), commodization is defined as a "dynamic process that erodes the competitive differentiation potential and consequently deteriorates the financial position of any organization" (p. 317). The authors claim that commodization frequently occurs through standardization, customer experience, and competitive imitation, which, in turn, lead to a profit squeeze. Profit squeeze can be described as deteriorating financial performance. Moreover, the authors claim that commodization is often considered a deliberate customer tactic that is targeted at increasing bargaining power by undermining or

Table 2.2 Strategies to counteract commodization

Value proposition base	Differentiation base
Product leadership	Product innovation and superior product qualities
Customer linking	Service innovation and customer bonding
Cost leadership	Operational excellence and fair value solutions

Adapted from Matthyssens and Vandenbempt (2008)

eliminating differences between offerings competing in the marketplace. Commodization can be considered a key challenge of business markets and, more importantly, it is also a main drive of servitization.

Decommodization is the reverse process which may take place when the commodization process leads companies to a profit squeeze, which is rather an unwanted situation. Therefore, certain strategies can be adopted in order to counteract commodization and target the mastering of certain niches. Matthyssens and Vandenbempt (2008) outline three main strategies based on different value propositions (Table 2.2)—(1) product leadership as a differentiator, (2) customer linkage as process integration media, and (3) cost leadership as a differentiation strategy focusing on efficient delivery of service.

2.4 Manufacturing and Service Organizations

In order to fully grasp the nature of servitization, it is also crucial to understand the generic division of organizations into manufacturing and service. The main characteristics of manufacturing and service organizations are summarized in Table 2.3, based on the research conducted by Reid and Sanders (2005).

Servitization implies a shift extending the offering beyond tangible products, which on the one hand enables organizations to achieve higher profit margins, while on the other posing additional challenges.

Servitization in manufacturing companies has been studied by several authors, mainly by using a case study approach. Some of the reported cases include:

1. Thales flight simulator systems and training of airline pilots (Davies 2004)
2. The Rolls-Royce "Power by the Hour" maintenance concept for the jet engines (Howells 2000)
3. Xerox copy machines with guaranteed unit cost per copy (Mont 2002)

A common component in all these examples seems to be bundling the maintenance component with the physical product and simultaneously putting the focus on life cycle costs and operational expenses. Organizational changes are needed and new types of contracts to clarify the responsibilities of each partner in imaginable situations. In addition to trust and contracts, technology is used to keep a close to real-time view on actual asset utilization.

Table 2.3 The main characteristics of manufacturing and service organizations

Manufacturing organizations	Service organizations
Product	
Tangible	Intangible
Storing and warehousing	
Possible	Not possible
Contact with customer	
Relatively low	Relatively high
Response time	
Relatively long	Relatively short
Requires	
Capital	Labor

Adapted from Reid and Sanders (2005)

2.5 New Service Development

According to Edvardsson (1997), there are three crucial service components: concepts, process, and systems. In service design or development, each of the aspects needs to be covered. The service concept is about packaging the offering to customers and standardization of the offering for service providing. The process factors refer to actual execution of the service transactions: both what is visible for the customer and what is not. The service component of the system is about the resources needed to conduct the process (Table 2.4).

The process of new service development may vary, depending on the company. Information about market demand may be collected systematically from advanced users, or companies may benchmark similar solutions from competitors or other industries. Kindström and Kowalkowski (2009) propose a four-stage framework for new service development. The stages proposed by the authors are summarized in Table 2.5.

2.5.1 Service Blueprinting

One possible approach to model service interaction from the process management point of view is to use the service blueprinting technique (Bitner et al. 2008). Service blueprinting is an approach reminding of the swimming lane diagram used in business process reengineering. The process flows from the top left corner toward the top right. Instead of having the resources and actors in columns, the layers are divided based on what the customer sees and what is not perceived directly. The idea of service blueprinting is focused upon the processes occurring in an organization while it interacts with its customers. Zeithaml et al. (2006) propose the following elements of service blueprinting:

- Physical elements needed to complete transactions.
- Customer actions in the service encounter.

Table 2.4 Service components

Component	Description
Service concept	What customer need and how to satisfy those needs (service offer)
Service process	Chain of local and central activities to produce the service
	Front office processes—those encountered by the customer
	Back office processes—internal, no interaction with customers
Service system	Resources required for service process and the service to be provided (company's organizational structure, physical and technical resources, customers, and employees)

Adapted from Edvardsson (1997)

Table 2.5 New service development (NSD) stages

Stage	Description
Market sensing	Occurring within a company, based on a dialogue with customers and any other actors such as systems integrators, consultants, and contractors
Development	Combining NSD and NPD efforts, more cross-functional and intraorganizational than NPD; processes are customer intensive
Sales	Tangible actions for commercializing and scaling up the new services. The main challenge of this phase is to help customers appreciate the distinctiveness and benefits of the new offering. Employees responsible for selling of the new offering should have sufficient knowledge. The organization must ensure the delivery of superior service
Delivery	Elementary different from the delivery of products. Services are created in an interaction with the customer during the process of delivery. Most services are people intensive and rich with interactions, which is why they often create new cost centers. Therefore, the delivery of services often requires specific service infrastructure

Adapted from Kindström and Kowalkowski (2009)

- Onstage contact employee actions, visible to the customer.
- Backstage contact employee actions which may be completed synchronously or asynchronously and are not visible to the full extent to the customer.
- Supporting processes help the main process and could also include information systems and other external key process elements.

Other notable concepts in this model include:

1. Line of interaction, which takes place between customer actions and onstage contact employee
2. Line of visibility, which takes place between onstage contact employee and backstage operations

In the context of service-dominant logic, service blueprinting as a method should be further enriched with mapping the customer's role in value cocreation. Service blueprinting is a visual method that allows seeing the overall idea of a situation. Figure 2.3 presents an example of a service blueprint for placing an order for repair

2.5 New Service Development

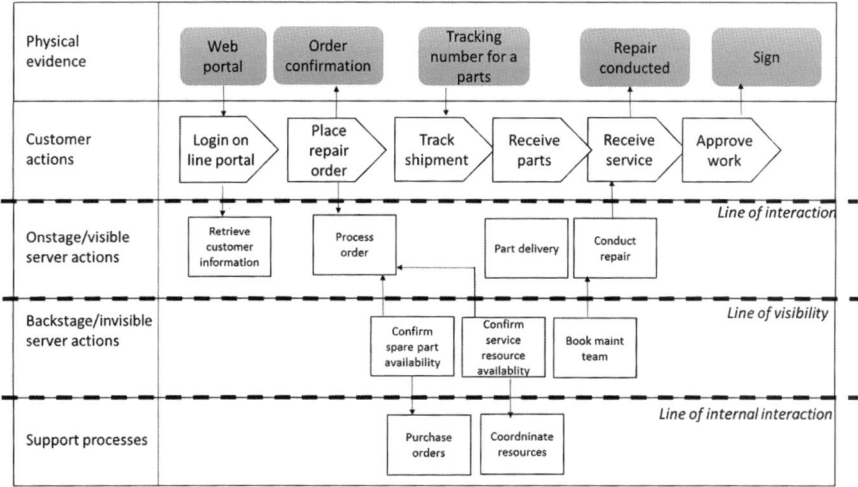

Fig. 2.3 An example of a service blueprint—ordering and receiving repair service for machinery

service by using an online portal. The physical and intangible service components of the delivery are intercoupled in this example and information is supplied throughout the process.

2.5.2 Service Innovation and Business Models

Service innovation and development of new business models are related with each other. The introduction of a new service product very often means a new business model in terms of pricing or service delivery. According to Kindström and Kowalkowski (2014), innovation seen from the service perspective means "a recombination of resources that creates new benefits for any actor—customer, developer, or others—in the business network" (p. 97). The authors claim that profitable manufacturers must be able to "capture an equitable share of the value created" (p. 97) (Fig. 2.4).

Osterwalder and Pigneur (2010) created a highly visual tool for generating business models—a business model canvas. Figure 2.5 presents a comparison of the elements of the generic business model canvas as proposed by Osterwalder and Pigneur (2010) (right-hand side) and the business model for service innovation as proposed by Kindström and Kowalkowski (2014) (left-hand side).

Kindström and Kowalkowski (2014) have proposed a systematic approach for development and analysis of business model for service innovation. Table 2.6 summarizes the elements of the model. The key elements are (1) offering, (2) revenue model, (3) deployment process, (4) sales, (5) delivery, and (6) customer relationships. The value network element is the supply chain delivering the services.

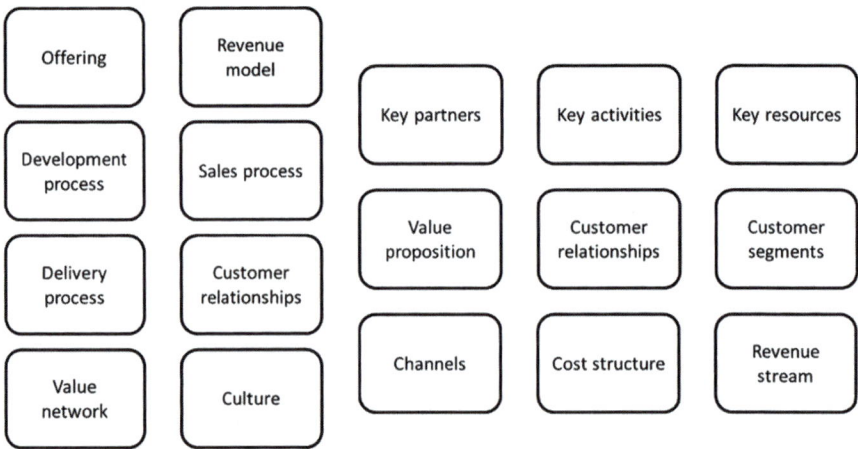

Fig. 2.4 A comparison of generic elements of business model canvas (adapted from Osterwalder and Pigneur 2010) and the business model for service innovation (adapted from Kindström and Kowalkowski 2014)

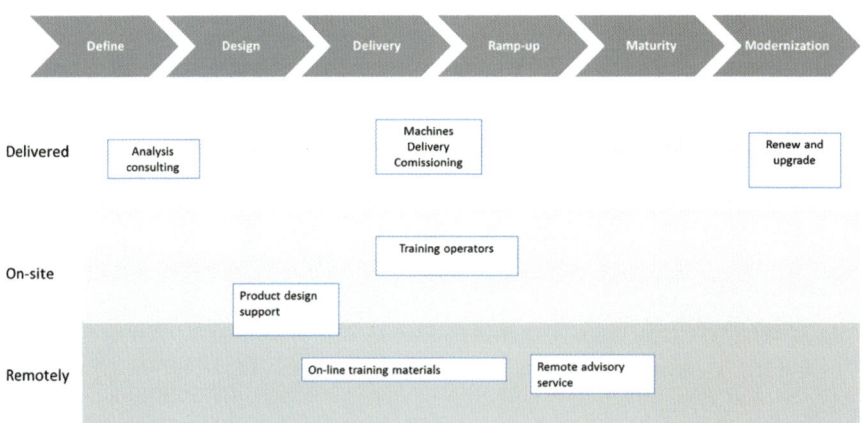

Fig. 2.5 System life cycle and service offerings

2.6 Product Life Cycle as Platform of Servitization

The life cycle of the product presents one approach to analyze possible service interactions with the customer. Maintenance services and extended warranties are typical initial steps toward the operation side. For machine operators, the life cycle analysis approach may be used as a reference model for development as well as an approach to use for total cost of ownership in decision-making.

2.6 Product Life Cycle as Platform of Servitization

Table 2.6 Business model for service innovations, adapted from Kindström and Kowalkowski (2014)

Component	Explanation
Offering	Understanding of what services to offer, as well as how to develop a complete portfolio of services is required as well as the agreement on the degree to which services would be standardized. Customer needing (Strandvik et al. 2012) must be aligned with supplier's offering
Revenue model	Product and process data as the key input for revenue models. Pricing capability is essential in determining how to charge for services and how to change service models if needed
Development process	Service development, sales, and delivery as the processes critical to service innovation
Sales process	Need to align incentive system with strategic service objectives in order to promote the sales of services and change the behavior of product-centric sales force. Customer involvement is critical
Delivery process	Successful service delivery requires an existing field service network. The process is an continuous customer-supplier relationship based on commitment and trust
Customer relationships	Stability of customer interaction as a facilitator of the development of lasting customer relationships. For instance, customers provide valuable information that enable suppliers to provide better services and improved customer satisfaction
	Customer embeddedness as an essential capability and it refers to an organization's ability to develop close
Value network	A distribution network is a meaningful resource in creating value as it not only provides service sales and delivery but also offers critical information about customers, service operations, and the market
Culture	Adding services might help organizations create internal awareness of the importance of services as well as the potential for adding more
	Creating a service culture requires a long-term orientation

2.6.1 Life Cycle for Machinery Delivery

The following example illustrates the business to business interactions in the case of acquiring and operating machinery (Fig. 2.6). The first phase is related to the definition of the customer needs and outlining a requirement specification. Analysis and consulting related to asset investment are conducted with the customer in this phase. Typical deliverables include reports, payback calculations, and simulation models. The next step in the process is design of the product as well as product configuration documents. Product specification and engineering services may take place in this phase. Also training for operators can be initiated in this phase. The actual physical delivery refers to logistics of the product and initiation of installation and commissioning work at the customer site. As machinery is becoming increasingly smart and integrated to the Internet, technology provides opportunities to utilize online delivery methods.

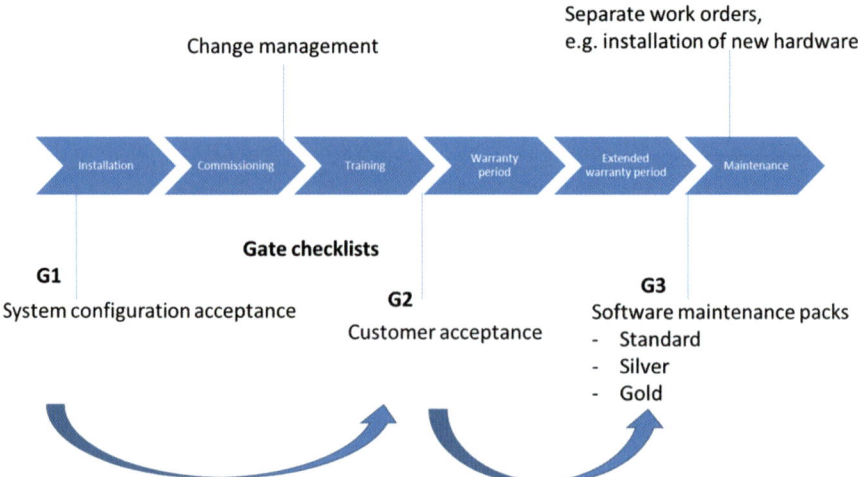

Fig. 2.6 Gate model in software delivery and transformation to maintenance

The training of operators may be conducted as an online service as well as presence education. Approval testing takes place and operations are phased to ramp up to expected volumes. In this stage, local support with physical presence is needed, and remote fleet management systems can be started to offer remote advisory service. Warranty is started and service level agreements are measured.

The final stage of the life cycle is approaching maturity. The maturity phase refers to full-scale utilization of the system and support for operations and maintenance is active. In this phase, operational needs are mapped and the system can be upgraded to support current needs.

2.6.2 Software Process Life Cycle

Many products have a software part which is responsible for creating an important set of functions and features of the system. Software enables the smartness of products. From the delivery point of view, it also brings flexibility and reduces delivery time. Software is a medium to adapt the same physical system to cover the various needs coming from customers. Software engineering is used to tailor the product to suit the customer needs (Fig. 2.6).

The challenge is that software is developed in cycles and consists of modular structures. Completely new features are developed in the research and development. Tasks include testing new technologies, prototyping and conceptual development, and managing requirements and issues coming as feedback and bugs from the customer sites. The core of the software, the kernel components used in all installa-

2.7 The Servitization Paradox   17

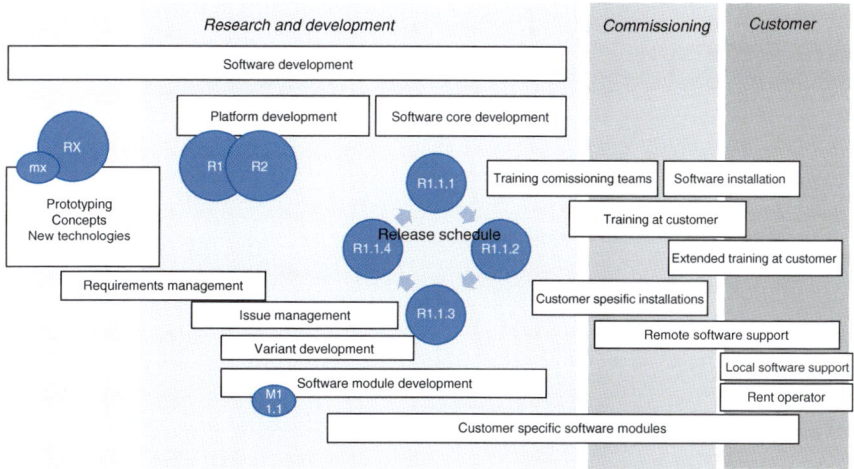

Fig. 2.7 Software life cycle and service offerings for a machine building company

tions, is produced by maintaining stability and backward compatibility. Smaller features and variants are developed based on requirements and issues. Organizations are keeping a standardized cycle to release software versions a few times per year. Hot fixes and patches are released in order to fix possible safety and security issues perceived.

The role of software organizations is multi-threaded by nature. On the one side, customer requirements are implemented as requested and general development of the software is performed. On the other side, installation and commissioning generates many tasks for software engineers to support the product-service delivery. Training support is needed and when an actual operator of the machinery gets started, the requirements are very often revised, which may result in a significant amount of rework. As the functionality of machinery is determined by the software, a continuous and frequent communication may take place for a long time after physical commissioning has been completed. A well-defined service level agreement and product packaging help in dialogue with the customer. It is important to know what is included in the deliveries, what is available as after-sales services, and what part needs to be agreed separately (Fig. 2.7).

2.7 The Servitization Paradox

The servitization process is complex and time consuming. Not all attempts are successful. Gebauer et al. (2005) define the service paradox as a situation when remarkable investments in service business expansion lead to increasing service offerings and higher costs while failing to gain the expected increased profitability.

In their research aimed at discovering the global trends in servitization, Neely et al. (2011) use longitudinal data from the years 2007, 2009, and 2011. The authors identify a widespread effort to servitize and highlight the importance of the servitization paradox based on numerous examples of unsuccessful servitization. The authors conclude that the success of servitization will be largely dependent upon building the right organizational capabilities and culture. Servitization offers many potential benefits for organizations. However, it is crucial to acknowledge the existence of the servitization or service paradox.

Chapter 3
Integrated Product-Service Systems

Gummesson (1995) was one of the first researchers to emphasize the fact that customers are not buying products any more but are rather buying offerings that are based on services which create an added value. The services offered need to solve real customer problems in an efficient and cost-effective way. The value creation of the service should be perceived instantly and be transparent to customers for evaluation of the performance. The transition from manufacturer to service provider can be justified not only from the perspective of potential financial benefits.

3.1 Types of Industrial Services

For machine-building companies, the range of services offered to the market can include a wide spectrum of offerings. In order to describe the possibilities, there are several ways to classify industrial services into categories. Edvardsson (1997) outlines eight types of industrial service offerings. Table 3.1 presents an overview of the types.

Repair and maintenance are examples of traditional services offered to customers. These require technical knowledge of the product and can help customers to extend the operational life cycle of the product. Training is an example of the support needed to ramp up operations and at the same time help to ensure the performance of the physical product. Once the installed base of the products is increasing and operational products are reaching the maturity phase of the life cycle, retrofit services may be offered to existing customers. Retrofit services can be used to modernize competitors' equipment to the customer. Process optimization requires skills related to actual operations rather than technical and manufacturing knowledge of the product. Service level agreements (SLA) can specify the offered service at each stage and condition. A high end SLA contract may include penalties and bonuses based on the performance delivered. Short-term rental generates flexibility and gives a financial component to service. Long-term rental has less flexibility but may provide a good financial solution packaged together with performance promise.

Table 3.1 Types of industrial service offerings

Repair services	Restoring capital equipment to sound condition after damage. Repair services comprise of corrective and preventive. Efficient information and communication technology are crucial to minimizing the effort needed to perform this type of service
Operations training	The training required before users have obtained the necessary skills to operate the equipment
Retrofit services	Replacing or adding of one or more hardware and/or software components to improve the overall performance or to minimize life cycle costs
Process optimization	Offering technical expertise to solve specific problems related to a customer's industrial production process
Safety inspection SLA	Comprises equipment inspection, functionality testing, and safety function testing. This service is difficult to fully standardize due to different safety regulations in different countries
High-end SLA	Entails preventive maintenance for a fixed price over a certain time period
Short-term rental	Customers rent for a fixed price per product, time, and/or usage of equipment for short-term emergency and temporary use
Long-term rental	The agreement is usually signed over several years. A complete solution including equipment, financing, maintenance, spare parts, and operations training. The customer agrees to cover the costs incurred by those activities

Adapted from Edvardsson (1997)

3.2 Product-Service System

Product-service systems (PSS) integrate aspects from the physical product side (goods) with an intangible service offering. According to Aurich et al. (2009), a product-service system comprises a goods-based product kernel which is supported by specific nonphysical services. The authors claim that the ultimate reason for adding a service component to a physical product is to increase the product performance expected by the customer throughout the complete product life cycle. Moreover, integrated product-service systems might significantly contribute to achieving customer satisfaction (Raja et al. 2013).

In order for organizations to be able to provide a complete product-service system (PSS), the careful integration of existing design and realization processes of both physical and nonphysical components is required. According to Aurich et al. (2009), the integration also addresses the process of configuring the product-service system.

PSS is often seen as a means for creating a competitive edge for customers, and according to Aurich et al. (2009), achieving competitive edge can be realized by careful configuration. Configuration often starts with a physical product, which is then enriched with specific nonphysical services, as presented in Fig. 3.1. In the case of industrial machinery, the physical component refers to the machinery delivered to the customer as well as supporting material such as spare parts. Software and information component refers to the automation side, user interfaces, and functionality of the machinery which is enabled by computerized software. This functionality can be updated in a cost-efficient way, and it enables improved performance,

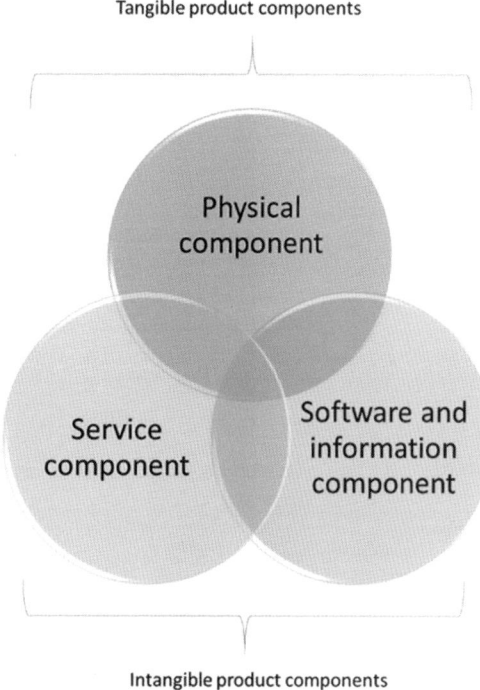

Fig. 3.1 A schematic view of product-service system

new functionality, or communication capabilities. As smart-connected products are becoming standard offerings in most fields, this component of the product is becoming increasingly important. Service component is the element of service offering – repairs, operations, support, training, and so on. An integrated product-service system requires elements from all of these categories.

Aurich et al. (2009) also call such an integrated approach a result-oriented one, while simultaneously claiming that the reverse approach is not typical of capital goods industries. Reim et al. (2015) highlight the need for multidimensional models that would identify and illustrate the various characteristics of PSS offerings while building upon the models already existing in the literature.

3.3 Benefits of PSS

The benefits of product-service systems are typically evaluated from the manufacturer's point of view. Servitization provides opportunities for manufacturers to operate closer to customers and to have a better understanding of the actual operational environment. This results in a stabilized revenue stream. However, the manufacturer or service provider is not the only perspective that can be analyzed. Mont (2002) outlines numerous benefits that stem from PSS. External service companies can do business

Table 3.2 Benefits of PSS

Manufacturing companies	Attaching additional value to a product
	Foundation for a growth strategy based on innovation in a mature industry
	Improved relationships with customers
	Improved total value for customers
Service companies	Extended and diversified service
	Safeguarding market share by adding a service component that is difficult to be copied
	Improved communication of product-service information
	Safeguarding a certain level of quality that is difficult to change (product quality)
Government/society	Supporting formulation of sustainable patterns of consumption
	New ways of understanding and influencing stakeholder relationships and viewing product networks – a basis for formulating new policies
	Supporting the creation of new jobs (as services are usually labor intensive)
Consumers	Greater diversity of choices available
	Added value through more customized offers of a higher quality
	Changing needs and conditions can be better satisfied by the flexible service component
	Consumers are relieved of responsibility for the product as it stays under the ownership of a producer through its entire life span
Environment	Decreasing the total amount of manufactured products by proposing alternative uses for those already existing
	Producer's responsibility for the product on the increase

Adapted from Mont (2002)

based on ecosystems provided by integrated product-service systems. Sustainability and employment are in the interest of governments and wider society. Consumers and the environment obtain benefits from both the value creation and sustainability aspects. Table 3.2 presents a summary of these benefits.

3.4 Characteristics of PSS

Based on the research conducted by Mont (2002), Table 3.3 provides an overview of the characteristics of PSS. The characteristics can be considered from four main perspectives: The first item is point-of-sale services focusing on the purchasing part of the process. The second item is the concept of product use, which refers to operations and value created by using the product. The third item is traditional maintenance service which prolongs the operational life cycle, and the fourth item concerns the revalorization services which focus on the final life cycle stages of the product.

A classic example of integrated product-service is the Rolls-Royce jet engine maintenance concept "Power by the Hour" which combines the reliability of the machinery with predictable maintenance.

3.5 Barriers to PSS

Table 3.3 Characteristics of PSS, adapted from Mont (2002)

Characteristic	Description
Point-of-sale services	Services such as personal assistance in shops, financial schemes for customers, explanations regarding products and their use, as well as marketing
Concepts of product use	Use oriented where the utility of a product is determined by the user
	Result oriented (utility provider determines product utility for the user)
Maintenance services	Product servicing aimed at extending the life span of a product, including maintenance and possible upgrades
Revalorization services	Services that aim at closing the material cycle of a product. This can be realized by, e.g., taking products back, reusing certain parts of new products, or recycling materials if reuse is not possible

Example: Case Rolls-Royce and "Power by the Hour" (Rolls-Royce Website 2014) "Power by the Hour" is a well-known service product concept developed for jet engine maintenance business. The concept has been developed and successfully applied from the 1960s. The idea of "Power by the Hour" is to offer a fixed cost maintenance program, which is based on engine hours. This approach makes it easy to estimate the expected life cycle costs of the engine. Rolls-Royce has trademarked the name, but similar concepts are commonly used by all jet engine vendors.

The business model is supported by engine health monitoring system allowing remote diagnostics. Replacement engines may be offered for off-wing repairs to minimize unproductive operations time.

The main benefit of the service is allowing operators to remove risk related to unscheduled maintenance events and make maintenance costs planned and predictable. The Rolls-Royce Website (2014) states: "The key feature of the program is that it undertakes to provide the operator with a fixed engine maintenance cost over an extended period of time. Operators are assured of an accurate cost projection and the cost associated with breakdowns."

The pricing is based on simple charging logic – per operational hour of the engine. The effect on the value chain is that by making operation of the jet engine predictable, the competition is shifted from investment costs to operational costs and reliability.

3.5 Barriers to PSS

Even though the shift toward product-service systems has many potential positive implications for both manufacturers and customers, certain barriers to product-service systems also need to be acknowledged. According to Mont (2002), the following issues present barriers to the introduction of product-service systems:

- Difficulties in developing scenarios for alternative use of products, as they often include elements situated at the interface between production and consumption

(sales), which calls for the involvement of different stakeholders in the process of designing both the product and service system.
- The challenge of identifying the existing social system or establishing a social system that would be needed in implementation of the PSS.
- The need for close cooperation with suppliers and service producers. Challenges inherent to integrated chain management (ICM) are also PSS. The sample challenges include the process of choosing the actors, sharing information transparently, and international restrictions in terms of material handling and regulations.
- The challenge of environmental impact. Multiple uses of the same products does not always lead to the decreased environmental impact.
- The challenge of changing systems and sources for gaining profit, the major problem being the changeover toward long-term orientation in financial planning.
- The inherent resistance of companies toward extending their involvement beyond the point of sale, as the involvement usually leads to increased manufacturer responsibilities for, e.g., the environmental impact of products.
- The requirement for a fundamental shift of an organization's corporate culture and commitment to market. This process requires both calendar time and resources to be allocated. Also the change of marketing concepts might cause the psychological barriers in companies to increase.
- Challenges in tracing the shift in service or manufacturing industries.
- The challenge of adding environmental aspects and considerations to the product development process, as it is usually seen as an unnecessary extending of the time to market.
- The challenge of customers not supporting ownerless consumption and the risk of not being able to predict consumers' reactions to the new PSS models.
- The growing complexity of customer demands, which are also increasingly difficult to predict.

3.6 Steps Toward Integrated Product-Service Systems

A transition from product manufacturer to service provider is challenging as it requires the creation of new organizational principles, structures, and processes. Additionally, new capabilities are needed and a business model changes from being transaction based to relationship based (Oliva and Kallenberg 2003). Oliva and Kallenberg (2003) propose a model for transitioning from a product manufacturer to a service provider, which consists of the following stages:

1. Consolidating product-related services is the initial step to harmonizing and packaging the service offerings.
2. Entering the installed base (IB) phase expands the scope of service supply to products that have already delivered to customers and are operational.

3. Expanding to relationship-based services and to process-centered services changes the focus from the product to operations – the actual use of the product and optimization of its use.
4. Taking over end use operations is the final and ultimate stage when customer transaction is based entirely on performed service.

Chapter 4
Strategic Improvements Through Industrial Services

Managing supply chains greatly depends on understanding the classification of industrial services, the customer expectation of the industrial services and how this expectation of the customer affects the quality of the services and demands for improvement, and also the flexibility that is demanded in the service relationship.

Companies are seeking new ways to win the competition. Products are often seen as similar by the customers and for this reason new differentiation methods are needed. This refers often to the ability of a company to adapt to changing environment and find ways to utilize its recourses (Barney 1991; Hunt and Morgan 1995; Hunt 1997). The perspective of industrial services as marketing and operations strategy allows to view both intangible and tangible assets as resources for creating value for chosen market segments (Hunt 1997).

These resources could be, for example, physical, human, organizational, informational, or relational (Barney 1991; Hunt 1997), and they may not always have an enabling capacity; for example, a resource may be strategically irrelevant for a firm or even reduce its efficiency or effectiveness (Barney 1991). In the present context, manufacturing firms use resources to operationalize their service strategies in order to gain a competitive advantage. In other words, resources are defined as having an enabling capacity. Different manufacturing companies focus on different resources to operationalize their service strategies, and research has demonstrated that many companies find it difficult to implement their strategies (Wouters 2004). The development of industrial services can also be realized in virtual manufacturing environments (Chen 2015).

4.1 Classification of Industrial Services

Categorization of consumer services has been well received by companies. Classification of industrial services seems to follow behind despite the certain similarities between these two areas. Common factors for both areas include the following properties:

1. Intangibility of the product.
2. Inseparability from the service provider and cocreation aspects.
3. Temporality – services are often created and consumed simultaneously.
4. Heterogeneity of the service delivery due to cocreation and variability.
5. Perishability of the service as inventory and storage concept does not really work with services.

For these reasons, it is also common that demand fluctuation affects the service provider's capacity utilization. Managing service capacity and performance related to time and cost has to operationalize with occasionally idle servers (Stanton et al. 1991).

Industrial services have been typically grouped into two categories, which are generic and wide by nature:

1. Maintenance and repair-related services – the service products in this category could include operations related to repair or equipment, cleaning services, and security.
2. Advisory services, which would include financial services, legal services, and engineering services, management consulting, for example (Kottler 1994).

Another approach to classify industrial services is to use the features of (1) replacement rate, which refers to frequency of use by the customer, and (2) essentiality, referring to the importance of conducting the service in the perspective of safety or reliable operation (Boyt and Harvey 1997).

According to Boyt and Harvey (1997), classification schemes are useful not only for explanatory reasons but, more importantly, the correct utilization of classification systems might significantly increase potential competitive advantage. Various classification schemes have been developed for goods and services, and with the advent of servitization, industrial services have also been classified. Table 4.1 presents an overview of existing systems as outlined and summarized by Buzacott (2000).

Buzacott (2000) highlights that the abovementioned classifications are not supported by appropriate theories. The classification proposed by Hayes and Wheelwright (1979) is, according to Buzacott (2000), both descriptive and prescriptive, while being useful as a diagnostic and design tool.

Buzacott (2000) proposes a model for structuring service systems that acknowledges the disturbances or variability that can stem from both inside and outside the system. According to Buzacott (2000), external variability occurs because the needs of individual customers are not known until the actual service occurs. Additionally, seasonal, cyclic, or random unexpected fluctuations in demand volume or service

4.1 Classification of Industrial Services

Table 4.1 An overview of the methods for classifying service systems

Reference	Criteria/explanation/dimensions used for classification
	Based on the degree/factor of
Chase (1978)	Tangibility vs. technology
	Tangibility vs. extent to which the service is provided
Shostack (1987)	Complexity of the structure of service delivery vs. degree of divergence allowed at each process step
Wemmerlöv (1990)	Divergence (service can be standard or customized)
	Customer contact (spectrum ranges from none to direct customer-service worker interaction)
Chase and Aquilano (1992)	Richness or information transfer of service delivery options
Silvestro et al. (1992)	Number of customers processed per day
	Balance between people and equipment, process vs. product balance

Adapted from Buzacott (2000)

product mix can take place. Internal variability, in turn, can arise from breakdowns in equipment, personnel absenteeism, or inadequate training. Buzacott (2000) claims that the main challenge of developing service systems is ensuring that customer demands can be met in an effective way. Moreover, the system has to tolerate its internal variability and uncertainty. Therefore, Buzacott (2000) proposes a model based on two dimensions—x and y dimensions on a chart. The x-axis represents the nature of the service offering. This axis presents the variability and complexity of the service operation. The y-axis presents the ability of the service providing system to cope with the uncertainty. According to Buzacott (2000), the optimal choice for structuring services is along the diagonal.

Buzacott (2000) proposes structuring service systems based on the following dimensions (y-axis):

- Specialization (workers specializing in a specific set of tasks)
- Parallel (any worker is able to perform all the tasks for any given customer)
- Series (multiple workers required to perform all the required tasks for a given customer)
- Bottom-up (based on the complexity of the diagnosis: if it increases with successive steps, the structure is bottom-up)
- Top-down (if the complexity of diagnosis decreases with successive steps, the structure is top-down)

The x-axis in the service system structure classification proposed by Buzacott (2000) refers to the variability and complexity of the tasks that need to be performed to fulfill the customers' requirements (Fig. 4.1).

Kowalkowski et al. (2011) have proposed a classification scheme for industrial service offerings based on the service focus (with a division into process-oriented and product-oriented services) and service scope (with a division into unbundled and bundled services) (Fig. 4.2).

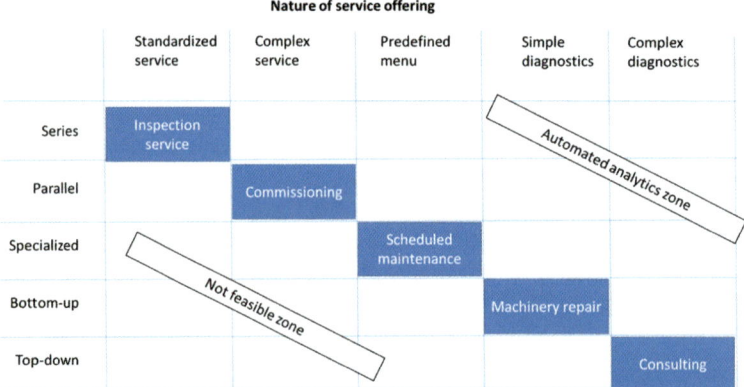

Fig. 4.1 Classification of service system structures. Adapted from Buzacott (2000)

Fig. 4.2 A classification scheme for industrial service offerings. Adapted from Kowalkowski et al. (2011)

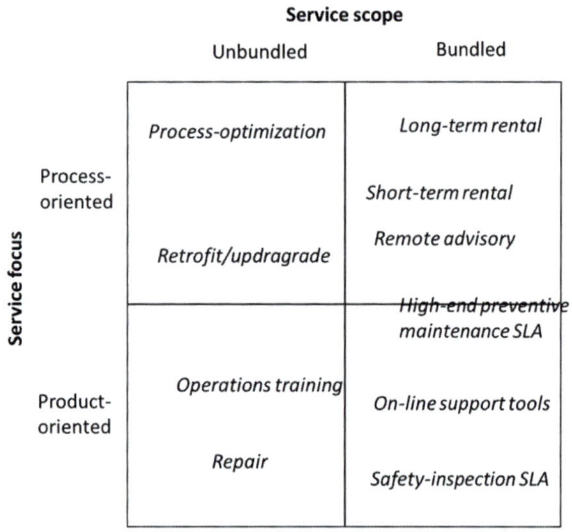

The classification can be done according to service delivery types (Fig. 4.3) as suggested by Schmenner (2009). The server facilities and processes vary according to labor intensity and required interaction levels. According to this quadrant model, the service types are service shop, service factory, professional service, and mass service.

A study by Partanen and Kohtamäki (2011) proposed a measurement method for the scope of industrial services, with a tool called SERVSCOPE. The tool was developed by analyzing the literature and consists of four dimensions: optimization services, research and development services, business services, and production information sharing services. From a theoretical point of view, it seems that the

4.1 Classification of Industrial Services

Fig. 4.3 Four types of service delivery. Adapted from Schmenner (2009)

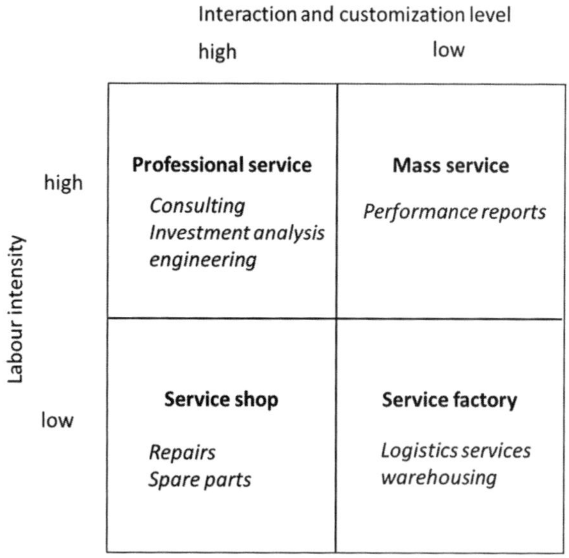

	Interaction and customization level	
	high	low
Labour intensity high	**Professional service** Consulting Investment analysis engineering	**Mass service** Performance reports
Labour intensity low	**Service shop** Repairs Spare parts	**Service factory** Logistics services warehousing

business analysis side of industrial services is developing in various areas where emerging technologies in IT and telecommunications could offer more advanced solutions. Novel approaches in supply chain practices could be developed on top of these.

If we analyze the situation from the product offering point of view, as proposed by Kohtamäki (Table 4.2), we can see some ladders of implementation – different service products correlated in different magnitudes. There are certain dependencies – some products are mandatory in different situations and others come later in the development chain. The same study also shows some interesting correlations among profit and sales-related constructs. As the data from Finnish companies shows, there is some knowledge on current offerings and the impact on profitability, but understanding of how business parameters can be achieved on the organizational and operational level is still limited.

Raddats and Kowalkowski (2014) call for reconceptualization of manufacturers' service strategies and identify the following categories of services: product attached, operation services on own products, and vendor-independent operation services. The authors outline three generic servitization strategies based upon the proposed categories:

- Service doubters (service is not a strong differentiator, therefore no focus on any particular service category)
- Service pragmatists (product-attached services as a key differentiator)
- Service enthusiasts (services are perceived as both a product differentiator and growth enabler)

Table 4.2 Literature on service business grouped into five aspects and service product types

Perspective	Service type	References
Research and development	Product engineering	Homburg et al. (2003), Samli et al. (1992)
	Prototype manufacturing	Kohtamäki et al. (2013)
	Feasibility studies	Homburg et al. (2003)
	Problem analyses	Oliva and Kallenberg (2003), Homburg et al. (2003)
	Manufacturability studies	Oliva and Kallenberg (2003)
	Research services	Homburg et al. (2003)
Product information sharing services	Product demonstrations	Homburg et al. (2003)
	Customer seminars	Homburg et al. (2003)
	User training	Oliva and Kallenberg (2003), Homburg et al. (2003), Morris and Davis (1992), Samli et al. (1993)
	Documentation	Kohtamäki et al. (2013)
	Written information material	Homburg et al. (2003)
	Customer consulting and support by phone	Homburg et al. (2003)
	Cost-benefit analysis	Homburg et al. (2003)
Optimization	Installation and commissioning	Oliva and Kallenberg (2003), Homburg et al. (2003), Morris and Davis (1992), Samli et al. (1993)
	Delivery service	Oliva and Kallenberg (2003), Homburg et al. (2003), Morris and Davis (1992)
	Technical support for similar products provided by other vendors	Homburg et al. (2003)
	Repair service	Oliva and Kallenberg (2003), Boyt and Harvey (1997a, b)
	Spare parts	Oliva and Kallenberg (2003)
	Electronic ordering system	Oliva and Kallenberg (2003), Homburg et al. (2003), Morris and Davis (1992)
	Recycling service	Oliva and Kallenberg (2003), Homburg et al. (2003)
	Product upgrade service	Homburg et al. (2003)
	Maintenance	Oliva and Kallenberg (2003), Homburg et al. (2003), Morris and Davis (1992), Samli et al. (1992)
	Warranty and extended warranty	Morris and Davis (1992)
Business	Procurement service	Homburg et al. (2003)
	Vendor-managed inventory service	Homburg et al. (2003)
	Mediation of products	Homburg et al. (2003)
	Project management	Homburg et al. (2003)
	Service for operating the product	Kohtamäki et al. (2013)
	Financing services	Homburg et al. (2003), Samli et al. (1992)
	Insurance services	Homburg et al. (2003)

Adapted from Kohtamäki et al. (2013)

Storbacka et al. (2013) propose utilization of a business model perspective to outline generic stages through which industrial companies are transforming toward solution-based models. The authors outline the following stages:

1. Customer embeddedness to integrate toward customer processes
2. Offering integratedness to integrate the product definition and specifications
3. Operational adaptiveness to develop the service delivery
4. Organizational networkedness to enhance the resource allocation and utilization within the service organization

By gradually following these stages, organizations are more likely to undergo a successful transformation that will last as it is based on alterations in business models.

Rabetino et al. (2015) and Buxel et al. (2015) propose the development of service offerings based on product life cycle, which is much better suited to the actual customer needs which might evolve over time, as well as product performance, which is also affected by time.

4.2 From Goods-Dominant to Service-Dominant Logic

According to Kowalkowski (2010), PSS offerings can also work as a means for creating value. Nevertheless, organizations should be aware of the transition process from goods-dominant to service-dominant logic, as the change requires time to occur. Kowalkowski (2010) suggests that in the case of manufacturing firms, more than just increasing the importance of a manufacturing's firm's product-service systems is implied. This means reframing the purpose of the organization as well as its collaborative role in value cocreation.

Kowalkowski (2010) also claims that the transition from goods-dominant (G-D) logic to service-dominant (S-D) logic can be seen as a two-dimensional occurrence. Therefore, manufacturing organizations might experience the processes of:

- Reflecting a change of strategic positioning of the manufacturing company in the market – this can occur through adding new services to the core offering.
- Reflecting new perspectives on value creation process with the customer.

Table 4.3 presents the transition process from goods-dominant to service-dominant logic.

Furthermore, Kowalkowski (2010) states that the most common misunderstanding that has arisen around the concepts of goods-dominant and service-dominant logic is the false assumption that the transition to service-dominant logic would mean always a focus purely on services while outsourcing manufacturing activities. The author outlines several managerial guidelines that service-dominant logic offers:

Table 4.3 Transition from goods-dominant to service-dominant logic

G-D logic	S-D logic
Offering a ready product	Customers create value on their own; assistance is provided
Producing value	Cocreating value
Customers are not interacting, treated as targets	Networks of customers, treated as resources
Focus on efficiency	Achieving efficiency through being efficient

Adapted from Kowalkowski (2010)

- Organizations should aim for transparency and symmetry of information in the exchange process. Customers are important collaborators, and therefore breeches of trust cannot occur.
- Organizations should aim at developing long-term relationships with customers.
- Organizations should view physical goods component of a product as transmitters of operant resources, while focusing on selling service flows.
- Organizations should support and invest in the development of specialized skills and knowledge that are foundational to the growth of business revenue.

4.3 Goods-Dominant Logic vs. Service-Dominant Logic

According to Vargo and Lusch (2008), there are two main perspectives when considering services. The first *goods-dominant logic* (also referred to as the *neoclassical economic research tradition* or *manufacturing logic*) can be described as more inclined toward *using principles developed to manage goods production to manage service production and delivery* (p. 255). The second logic, as outlined by Vargo and Lusch (2008), is *service-dominant logic*. According to Vargo and Lusch (2008), service-dominant logic *provides a service-based foundation centered on service-driven principles* (p. 255).

Vargo et al. (2008) draw the main difference between goods-dominant and service-dominant logic on the basis of how value is perceived. The authors claim that the main difference lies in the fact that goods-dominant logic defines value as created by the manufacturer, while service-dominant logic sees value as always being cocreated by customers. Vargo et al. (2008) outline the following *foundational premises* of service-dominant logic:

- Service product is the fundamental basis of exchange.
- Indirect exchange between the customer and vendor is the fundamental basis for exchange.
- Service provision distribution system is based on physical product.
- Operant resources are the fundamental source of competitive advantage.
- All economies are service economies.

4.3 Goods-Dominant Logic vs. Service-Dominant Logic

Table 4.4 Comparison of goods-dominant and service-dominant logic

Criteria	Goods-dominant logic	Service-dominant logic
Value driver	Value in exchange	Value in use or value in context
Creator of value	Focal organization and other organizations within supply chain	Organization, its partners, and customers networked
Value creation process	Value embedded in goods or services and added by enhancing or increasing attributes	Value proposed through market offerings; through use customers continue the process of creating and adding value
Purpose of value	Increasing profitability of an organization	Increasing adaptability, survivability, and the overall resilience of the system
Measurement of value	The amount of nominal value, price received in exchange	Adaptability and survivability of the beneficiary system
Role of firm	Producing and distributing value	Proposing and cocreating value, providing service
Role of goods	Units of output, operand resources that are embedded with value	Vehicle for operant resources, enables access to the benefits of firm competences
Role of customers	To use or destroy the value readily created by the provider	Cocreation of value through integrating firm-provided resources with other private and public resources

Adapted from Vargo et al. (2008)

- The customer is always cocreating value.
- An enterprise itself is not able to deliver value, but it is offering value propositions.
- Orientation on customers and relations with them.
- All social and economic actors can be perceived as resource integrators.
- Value is always defined and detected by the customer.

Vargo et al. (2008) provide a comprehensive comparison of goods-dominant and service-dominant logic. Table 4.4 is outlined based on the research conducted by Vargo et al. (2008).

Chapter 5
Improving Marketing and Operations Strategy Through Industrial Services

Industrial services can be used to support the marketing and operations strategy of an industrial firm. The possibilities include using the coproduction aspect in customer relationship, transforming the business toward service-dominant logic, using service offering and delivery as part of global expansion strategies, and redesigning the service delivery by analyzing and developing the service supply chain structure.

5.1 Coproduction of Service Offering and Customer Experience Management

According to Lusch et al. (2007), the coproduction of services generally requires the increased involvement of the customer. Table 5.1 presents an overview of the factors affecting the coproduction of service offerings. The outlined factors also explain how much a customer wishes to be a part of service operations.

According to Lusch et al. (2007), the fact of customers being increasingly involved in the process of coproduction of service offering leads to an increase in instances of customer contact or touch points which can be treated as a basis for managing customer experiences.

Schmitt (2003, p. 17) defines customer experience management as "the process of strategically managing a customer's entire experience with a product or a company." The author proposes the following elements of the customer experience management:

- Analysis of the experiential world of the customer
- Building the experiential platform
- Designing the brand experience
- Structuring the customer interface
- Engaging in continuous innovation

Table 5.1 Factors affecting coproduction of service offerings

Factor	Explanation
Expertise	If a customer has the proposed expertise, their involvement in service coproduction is more likely to occur
Control	Coproduction is more likely to occur in a situation when a person is interested in exercising control over the process or the outcome of a service
Physical capital	Coproduction is more likely to occur if the party has the required physical capital
Risk taking	Coproduction involves physical, psychological, and/or social risk taking. However, risks do not have to directly increase with coproduction since coproduction also has the potential to reduce risks
Psychic benefits	Pure enjoyment is one of the main reasons why people engage in the process of coproduction. Enjoyment can be classified as one of the psychic (experiential) benefits
Economic benefits	Participation in the coproduction of services might be explained from the perspective of the perceived economic benefits

Adapted from Lush et al. (2007)

5.2 Increasing Competitiveness Through Service-Dominant Logic

The adoption of service-dominant logic can also be seen as a means of building and increasing the competitiveness of an organization. Table 5.2 presents an overview of the most important aspects and means by which this can be achieved. Typically, the reasons for an organization to implement service-dominant logic fall into several categories at the same time. Market demands and competition go hand in hand. Sometimes closer proximity to the customer and changing position in the value chain are needed in order to survive in an era of fast-advancing technological progress.

5.3 Global Service Strategies: Service as a Means of Expanding Business

With increasing competition and modern information and communication technology, organizations are not just encouraged, but often forced, to make their services global in order to stay competitive. Moreover, governments are gradually acknowledging the global market changes as well as the benefits of global services and beginning to offer incentives for organizations that wish to globalize their services. McLaughlin and Fitzsimmons (1996) propose several strategies that organizations can adapt in order to make their service offerings global. The proposed strategies are based on several globalization factors, such as:

- Customization (front-/backroom oriented)
- Customer contact (front-room operation based)

5.3 Global Service Strategies: Service as a Means of Expanding Business

Table 5.2 Elements of competitiveness and means of achieving it

Elements of competitiveness	Explanation/means of achieving
Service as a basis for competition	One of the premises of service-dominant logic is that it is not the products that are the ultimate aim of the customer's acquisition but rather the benefits available through the service provider—customers are purchasing solutions.
Knowledge, collaboration, and sustainable competitive advantage	Knowledge rather than service is the primary source of competitive advantage (competences as a source of competitive advantage).
Dynamic value proposition	In a dynamic environment of constant change, organizations are unable to become static in their value proposition—innovation in services is dependent upon the collection of competences.
Collaborative competence	Critical in complex, dynamic, and turbulent environments.
Absorptive competence	The ability of an organization to draw upon important trends and know-how from the external environment, it is also useful in absorbing new information from partners. This competence helps organizations in the process of transforming external environments into important sources of resources.
Adaptive competence	The ability of an organization to adjust to changing circumstances.
Building competitive advantage	By coproducing and cocreating value through collaboration.

Adapted from Lush et al. (2007)

- Complexity
- Information intensity (information can be globalized once it is digitalized)
- Cultural adaptation (adapting services that would sell in the customer's home country)
- Labor intensity (organizations are usually seeking a less expensive and well-educated workforce)

According to McLaughlin and Fitzsimmons (1996), there are five basic service globalization strategies (Table 5.3):

- Multi-country expansion (exporting a successful service to another country without modifications or with minor ones, for example, McDonald's)
- Importing customers (customers come to the location of the service because of its unique features, for example, Disneyland)
- Following your customers (services that follow previously global customers around the world)
- Service unbundling (breaking up service components in order to determine which ones can be contracted out)
- Beating the clock (gaining competitive advantage by bypassing the constraints of domestic time zones)

Table 5.3 Global service strategies

Globalization factors	Multi-country expansion	Importing customers	Following customers	Service offshoring	Beating the clock
Customization	A standardized service	Strategic opportunity	Re-prototype locally	Quality and coordination	More need for reliability and coordination
Complexity	Low level, routines	Strategic opportunity	Modify operations	Opportunity to focus	Time compression
Information intensity	Satellite networks	On-site advantage	Transfer experienced experts	Training	Exploit opportunity
Cultural adaptation	Modify service	Accommodate foreign guests	Possibly needed	Cultural understanding	Common language needed
Customer contact	Train local labor	Develop foreign language and culture knowledge	Develop foreign customers	Back office process development	Provide extended hours of service
Labor intensity	Reduce labor cost	Increased labor costs	Local service personnel	Reduce labor costs	Reduced labor costs
Others	Government restrictions	Logistics	Infrastructure needed	Communication between home office and service desks	Capital investments

Adapted from McLaughlin and Fitzsimmons (1996)

5.4 Service Supply Chain Structure

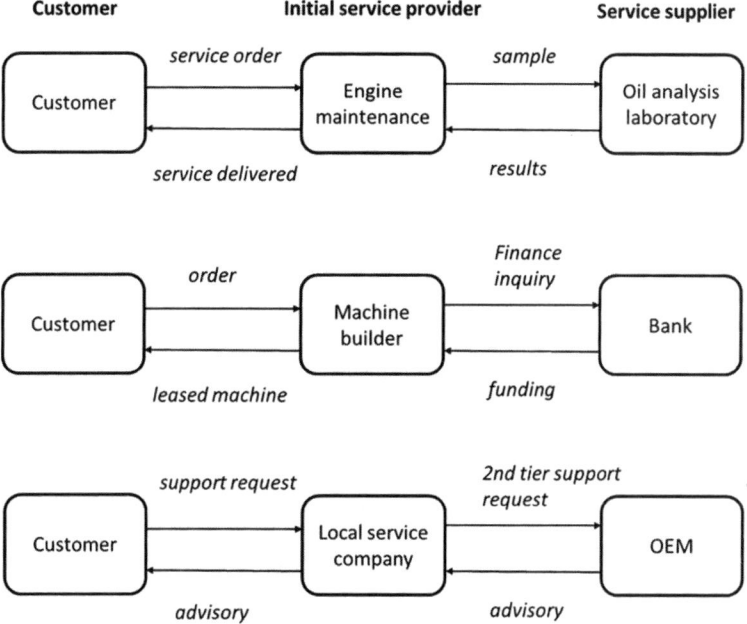

Fig. 5.1 Bidirectional supply chain examples

5.4 Service Supply Chain Structure

Service supply chains are related to physical supply chains in many ways. Supply chain actors are connected with order-delivery transactions, and process performance may be assessed from a system view rather than from a single actor level (see Prasad and Selven 2010; Löfberg et al. 2010; Avery and Swafford 2011; Boonitt and Pongpanarat 2011; Edvardsson et al. 2008; Ivens 2005). Services form supply chains in the same way as physical products—such as a maintenance engineer for engines taking oil samples and sending them to a laboratory and then the process being completed after the results of the laboratory tests are received. Supply chain connections may be bidirectional and form a loop, causing the customer waiting time in the process (Fig. 5.1).

Managing capacity is important in both cases as the flow of orders depends on resources. Service supply chains typically also include some physical products needed to complete the process. For example, the maintenance of a machine requires personnel to carry out the work as well as parts and tools to complete the task. However, as physical products can be stored in the inventory, this is not the case with services.

Service is coproduced at the time of consumption. Managing the capacity of servers due to the perishable nature of the process requires flexible time allocation for the workforce or the possibility to adjust demand. Some possible strategies to manage productive capacity include:

1. Transfer of knowledge to customers by using web-based access to FAQ banks
2. Substituting technology for service personnel
3. Educating customers to complete parts of the service themselves

All these approaches may have an impact on customer experience, and using these strategies within service product offering should be planned as part of cocreation process.

Chapter 6
Service Delivery

Service delivery is the operational part of the service and based on the supply chain consisting of processes and actors. Service delivery is designed around service provider, customer, suppliers, and the technology employed.

6.1 Service Delivery Concept

Transformation from products to services is a process that should be designed well. Service product positioning and pricing are important specifications. Delivery of the service is still the most important part as it defines the interaction of the service encounter. For this reason, the delivery of services should be well planned and part of a wider-reaching system design. Table 6.1 presents the most important definitions of the service concept that have been outlined by various researchers over the years.

According to Ponsignon et al. (2011), there should be an alignment between business strategy, service concept, and the design of a service delivery system. This principle is often described as a service strategy triad that draws special attention to the need for an integrated service design. This implies that the service delivery must support the realization of the service concept. Figure 6.1 illustrates the concept of a service strategy triad consisting of market, concept, and delivery system elements (Table 6.2).

6.2 Service Delivery System Design

The service delivery system design should answer a question of how the service concept is delivered to customers. According to Ponsignon et al. (2011), the most important factors to be considered while designing a service delivery system are the following:

Table 6.1 Service delivery and definitions of the concept

Reference	Definition
Sasser et al. (1978)	Goods and services bundled together and sold to the customer
Karwan and Markland (2006)	Tangible and intangible elements composed into a package
Collier (1994)	Conveys the benefits and value proposed to customers
Roth and Menor (2003)	A company's value proposition which is represented by varying degrees of customization and, therefore, requires various service delivery systems

Adapted from Ponsignon et al. (2011)

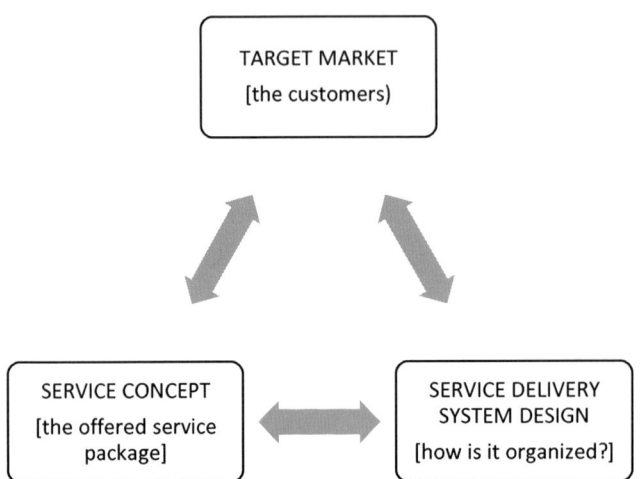

Fig. 6.1 The service strategy triad. Adapted from Ponsignon et al. (2011)

- Role of people in the system
- Role of technology, tools, and equipment in the system
- Role of location and layout
- The service venue – service facility and the processes through which services are delivered
- Jointly planned aspects of the dimensions of people, service processes, and physical elements

The design process itself has several characteristics recognized in earlier studies. Table 6.3 presents a summary of service delivery system design characteristics.

6.3 Customers' Roles in Service Delivery

Customers participate in the service encounter always to some extent. Coproduction concept has been used to describe customer participation in the event. According to Bitner et al. (1997), there are three types of customer participation in service experiences:

Table 6.2 Service delivery methods

Service offering	Local organization role	Centralized organization role	Online, remote managed service
Repair	Customer interaction	Concept development, ICT applications, product information	Software upgrades
	Completion of service		Remote advisory services
Operations training	Customer interaction	Concept development, product information	Webinar materials, online training packages
	Completion of service		
Retrofit	Lead generator for sales	Development of offering, resources, performing designs	Support for ramp-up
	Commissioning		Remote advisory services
Process optimization	Lead generator for sales	Development of offering, resources, performing designs	Online analytics service
	Service performing		
Safety inspection SLA	Customer interaction	Concept development, standardization, contracts, support for local organizations	Monitoring performance
	Lead generator for sales		
High-end SLA	Customer interaction	Concept development, standardization, contracts, support for local organizations, risk management	Monitoring performance
	Lead generator for sales		
	Service performing		
	Local development		
Short-term rental	Customer interaction	Concept development, standardization, contracts, support for local organizations	Asset utilization analysis for product vendor
	Lead generator for sales		
	Service performing		
Long-term rental	Customer interaction	Concept development, standardization, contracts, finance, support for local organizations	Total cost of ownership analyses
	Lead generator for sales		
	Service performing		

Source: Adapted from Kowalkowski et al. (2011)

- Customer as productive resource
- Customer as contributor to quality, satisfaction, and value
- Customers as competitors

The types of participation can be described from the perspective of the influence that they potentially have upon the service delivery process. Table 6.4 presents the influence of customers and their roles in the service delivery processes.

6.4 Customer Expectation of Industrial Services

Customer expectation is a central focus area in consumer-related services. Customer service has been pointed out to be the most important competitive factor in industrial markets (Clark 1993).

Table 6.3 Service delivery system design characteristics

Service delivery system	Professional service	Service shop	Service factory	References
Role of people				
Skills needed	High	Low	Medium	Kellog and Nie (1995), Silvestro (1999), Chase and Tansik (1983)
Degree of employee discretion	High	Medium	Low	Silvestro et al. (1992), Buzacott (2000), Lovelock (1983)
Role of technology				
Level of routineness	Low	Medium	High	Wemmerloev (1990), Buzacott (2000)
Level of automation	Low	Medium	High	Kellog and Nie (1995), Schmenner (2009), Silvestro et al. (1992)
Role of location and layout				
Location	Distributed (close to customer)	N/A	Centralized	Chase and Tansik (1983), Kellog and Nie (1995), Wemmerloev (1990), Cohen et al. (2000)
FO-BO configurations	Service oriented	N/A	Efficiency oriented	Metters and Vargas (2000)

Source: Ponsignon et al. (2011), adapted from Johansson and Olhager (2004)

Table 6.4 Customers' roles and their influence upon service delivery

Customers' role	Influence upon service delivery
Customers as productive resources	Customers aid service delivery if they are viewed as partial employees, self-service as the most extreme case.
Customers as contributors to quality, satisfaction, and value	Customer participation might stem from pure enjoyment, discounted price, or willingness to influence or control the service outcome.
Customers as competitors	A decision on whether to produce services for themselves (internal exchange) versus having someone providing the service to customers (external exchange) is crucial as it determines whether the service will be delivered or not. In that sense, customers compete with service providers.

Adapted from Bitner et al. (1997)

For industrial services these features are present as well and defined by customer segments. Understanding the importance for each case is crucial. For those services where personal involvement is not critical and possibility to replace the service is a feasible option, the expectations are typically low. On the other side, customer expectations are high in services involving personal communication which are closely related to successful operations of physical goods, and replacing the service is difficult. Service interaction is typically conducted by different actors. The roles may typically include technical staff who are performing the actual service delivery, supporting functions related to logistics and physical deliveries; administrative staff related to order intake, reporting, and invoicing; and experts related to process optimization and high-level advisory support.

Customer expectation is always compared with the actual service delivery. Service quality is the perceived difference between these two parameters. The objective of customer service function is to provide customer satisfaction and by doing so keep customers loyal to the vendor.

Chapter 7
Managing Service Delivery

Managing service delivery process requires a seamless flow from specifications (service-level agreements, SLAs) to operational performance metrics including the aspects of quality and flexibility. Coproduction challenges the delivery process as customer may own and control the physical asset. Managing the installed base and assets which are customer's or third party's property is a typical feature of advanced industrial services.

7.1 Service-Level Agreements

A service-level agreement can be described as a contract between a service provider and end user. This specification defines the service offered and to some extent the process delivering the service. A service-level agreement is output based, which means that end user defines the level of service that is expected from the service provider side.

Paschke and Schnappinger-Gerull (2006) define SLAs in terms of a document that describes the performance criteria a provider promises to meet while delivering a service. Moreover, SLAs typically include remedial actions and possible penalties that will take effect if performance fails to meet the promised standard.

Levels of service agreements are defined through metrics (KPIs, key performance indicators) and should be outlined in a way that the following are addressed. On the general level, SLAs should cover the aspects of volume and quality of work (including precision and accuracy), speed, responsiveness, and efficiency. Based on the work by Paschke and Schnappinger-Gerull (2006), the following table summarizes the essential aspects of SLAs (Table 7.1).

The exact scope of SLA will vary, depending on the nature of service; however, on a general level, the following aspects usually need to be addressed and measured (Beaumont 2006):

Table 7.1 Essential aspects of SLAs

Element	Definition
SLA metrics	Measure performance characteristics of service objects.
SLA rules (service levels and guarantees)	Representation of promises and guarantees, can be represented in the form of "if-then" rules.
IT measurement process	Used to define common practices such as incident, problem, configuration, change, or service-level management.

Adapted from Paschke and Schnappinger-Gerull (2006)

- Description of a service being provided
- Reliability (when the service is available, e.g., percentage uptime)
- Responsiveness (the punctuality of services to be performed in response to requests)
- Procedure for reporting problems (who should be contacted, which problems, what steps should be taken to resolve problems quickly and efficiently)
- Monitoring and reporting the service level (who will monitor performance, what kind of data will be collected, how often, how much access to performance statistics will be given to customers)
- Consequences for not meeting service obligations (how to proceed in the case of failure to comply with SLA; customers might be given the right to terminate a service relationship)
- Escape clauses or constraints (circumstances under which the SLA does not apply)

According to Marilly et al. (2002), the following aspects should be addressed in SLAs:

- The responsibilities of customer and service provider
- Procedures to be invoked in the case of violation of SLA's guarantees
- Service pricing and discounting mechanisms to be put into force when SLA commitments are not satisfied
- Service description and the commitment to quality of service
- Quality of service metrics and corresponding thresholds that must be guaranteed by the service provider
- Methods for measuring service performance, period during which measurement will be performed, and method and frequency of reporting
- Service schedule
- Reporting to customer (on the quality of services delivered)

The process may be divided into phases according to life cycle. An example of this approach is presented by Marilly et al. (2002) who outline the following stages of an SLA life cycle (Fig. 7.1) as follows:

- Product/service development (identification of customer needs and network capacities, preparation of service templates)
- Negotiation and sales (SLA is negotiated with customer; resources are reserved)

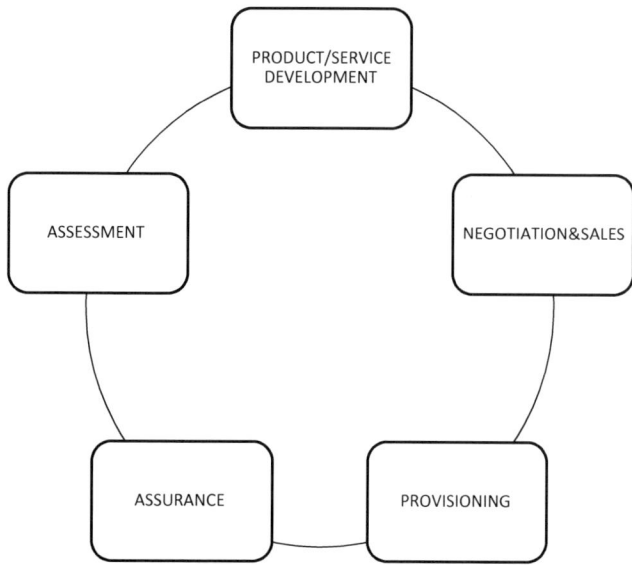

Fig. 7.1 SLA life cycle. Adapted from Marilly et al. (2002)

- Provisioning (resource provisioning and service activation)
- Assurance (SLA is monitored and reported, detection and handling of SLA violations)
- Assessment (assessment with the customer and internal operator assessment)

7.2 Performance Measurement

Key performance indicators (KPIs) as business metrics used to evaluate and manage performance are an important tool for any organization. Standardized frameworks for various areas of organizations have been presented. For physical delivery of a product, key performance indicators are typically related to cost of delivery, response time, reliability of delivery, and increasingly sustainability aspects such as emissions or environmental footprint. The Supply Chain Operations Reference (SCOR) model is a good example of a standardized approach to key performance measurement in this perspective. It takes into account all these parameters and suggests metrics for both high level and more detailed shop floor level.

Key performance indicators for service delivery are different. Internal efficiency and response to customer needs are important features. Taking into account the perishable nature of service and coproduction aspect, the implications have differences. Common metrics are set in agreements, which typically outline:

1. Performance of the asset in operation—operational efficiency and the reliability metrics of using an asset
2. Number of service requests per month agreed as part of the contract
3. Guaranteed resolution time for requests

On the detailed level of operations, the metrics could include the following measures:

1. Volume and categories of service requests
2. Service request fulfilled
3. Average response time to complete service request
4. Resource utilization

These aspects may conflict with each other. Nevertheless, it is evident that focus on a higher level is closer to asset management and on a detailed operational level will concern the internal resourcing decisions. Connecting asset management key performance metrics to drive the operations will require more understanding of customer asset utilization and customer business than efficient operations of maintenance.

7.2.1 Service Delivery Quality

Managing the quality of service adds some complexity for companies that have built quality systems to support physical products and deliveries. As services are intangible by nature and are cocreated with customers, the quality measurement is subjective. A widely accepted quality management framework for services is Parasuraman et al.'s (1991) SERVQUAL framework, which analyzes the gaps between service expectations and actual perceived service delivery. The five main dimensions in this framework are:

1. Tangibles, which refer to physical facilities, appearance, and equipment
2. Reliability, referring to the ability to perform the promised service dependably and accurately
3. Responsiveness, which may be seen as the ability to help customers and provide service quickly when needed
4. Assurance—the skill, knowledge, and capabilities of service delivery personnel to produce confidence in the process
5. Empathy—understanding customer needs and paying attention to solving customer problems

Each aspect in the SERVQUAL framework is measured by using a survey instrument, and gap analysis is conducted to perceive any service quality gap (see Fig. 7.2). The service quality gap refers to the difference between a customer's expectation and the perceptions of the service delivery. More recent developments in the field of measuring service quality propose analysis through the lens of complaints and praise (Vukelja and Runje 2014).

7.3 Installed Base Management

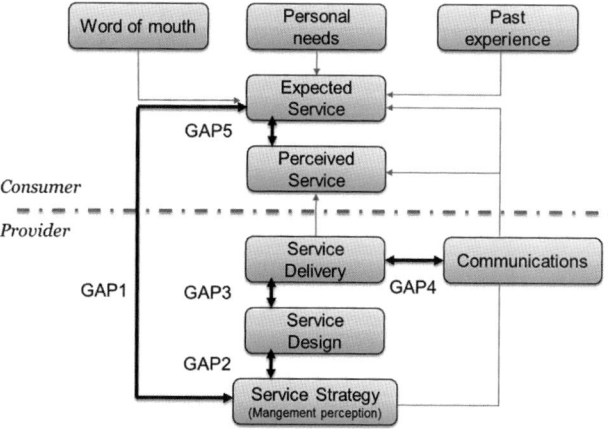

Fig. 7.2 Gap analysis of service quality. Adapted from Parasuraman et al. (1991)

7.3 Installed Base Management

According to Krikke et al. (2004), installed base refers to the total number of delivered and installed units of a certain product in the market segment. Activities related to management of installed base include service operations such as replenishment, repairs, predictive maintenance, spare part logistics, software patches, and generic system upgrades. Having an access to installed base of products in the market offers a valuable view to actual operational use of the products. This visibility can enable smart service products that can add value for the customer by enhancing the life cycle of the product.

According to Robotis et al. (2012), installed base management is defined as "the policy in which the manufacturer leases the product to customers, and bundles repair and maintenance services along with the product" (p. 236).

In installed base management, profit is sometimes generated through installations sold at competitive price and then compensated for through high margins on life cycle services such as spare parts and maintenance. The company builds its customer database through relatively low costs of installations, which implies enlarging the market for potential services. However, companies pursuing installed base management need to be able to build and maintain a competitive lead to keep the high margin levels (Van Looy et al. 2003).

Installed base management is widely used in sectors such as IT (Xerox), heavy machinery (Caterpillar), elevators (Otis), and transportation (Scania). Installed base management requires access to operational information: when product was installed and commissioned and where is the current location of product. Many systems go beyond this toward operational and maintenance information.

The pricing contracts for installed base management can be outlined on the basis of the following factors (Robotis et al. 2012):

- The amount of products already sold (this provides lower costs from remanufacturing some future products)
- The duration of a product leased to the customer (will the lease be of the same duration for all customers?)
- Single tariff versus multiple tariff structure (depending on the duration of the lease service)

Robotis et al. (2012) outline the following conclusions regarding installed base management, based on a comprehensive literature review performed in the field of marketing and operations management:

- Creation of network externalities that set industry standards based on different types of compatibility following innovation.
- Consumers should be given incentives to join the installed base to create momentum, as this eventually leads to increased utility from the complementarity of products and services, as well as through increased network effects.
- Price-skimming strategy is the most optimal for increased word of mouth and gives customers the incentive to buy future products.
- Installed base management provides more benefits to a firm in a competitive environment rather than in a monopoly.
- The service from the manufacturer and the independent service provider will be maximally differentiated in service quality and price, with the manufacturer offering the highest quality.
- The coordination of production structure, collection rate, and component durability is essential for the maximization of production and cost savings from remanufacturing.

Oliva and Kallenberg (2003) outline several definitions of installed base management that facilitate understanding of the concept. The authors emphasize the following aspects:

- Services are not restricted to those bundled with the products; installed base management encompasses all services that are required by the end user to obtain a desired functionality.
- Service suppliers also compete in the installed base management market—they are not restricted to product manufacturers, component manufacturers, system integrators, or end users' maintenance units and third parties.
- End users are not limited to industrial firms.

Oliva and Kallenberg (2003) also refer to certain benefits that stem from installed base management. Oliva and Kallenberg (2003) also state that by integrating the value chain from product design to service provider, product manufacturers are able to reap the following benefits:

- Lower customer acquisition costs—manufacturers are involved in the sales of new products, which means that they have information regarding the new equipment acquired to the installed base.

Table 7.2 Product-oriented services

Services oriented on product	Services oriented on end user's process
Transaction based	
Basic installed base services	Professional services
Documentation	Process-oriented engineering
Transport to client	Process-oriented R&D
Installation and commissioning	Spare part management
Product-oriented training	Process- and business-oriented training
Customer service/help desk	Process- and business-oriented consulting
Inspection/diagnosis	
Repairs/spare parts	
Product updates/upgrades	
Refurbishing	
Recycling/machine brokering	
Relationship based	
Maintenance services	Operational services
Preventive maintenance	Managing maintenance function
Condition monitoring	Managing operations
Spare part management	
Full maintenance contracts	

Adapted from Oliva and Kallenberg (2003)

- Lower knowledge acquisition cost—the product manufacturer has already acquired knowledge on the product-service requirements over the life cycle of the product.
- Lower capital requirements, as manufacturers already are in possession of specialized production techniques that are necessary for spare part fabrication or for upgrading existing equipment.

Oliva and Kallenberg (2003) divide product-oriented services into two main categories, as presented in Table 7.2.

7.4 Enterprise Asset Management

Enterprise asset management refers to the life cycle management of assets of an organization. This kind of systematic approach has been developed in industries which have high reliability requirements and where the cost of failure would be high. Oil and gas, marine, and process industries have developed standardization and tools to manage important infrastructure.

Enterprise asset management approaches such as standards on ISO 55000 outline systematic approaches how to manage operations and maintenance of large

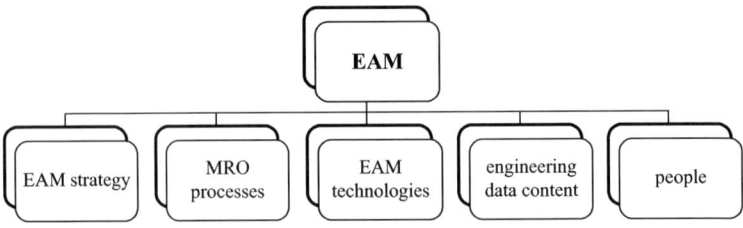

Fig. 7.3 Building blocks of enterprise asset management (Adapted from MacArthur (n.d.))

Table 7.3 EAM elements and their functions

EAM element	Explanation	Function
EAM strategy	How to produce the highest capacity at the lowest cost	Sets the overall direction of development
Maintenance, repair, and overhead (MRO) processes	A variety of processes encompassing asset management and maintenance function	Supports the achievement of the goals set by EAM strategy; processes must be engineered to the highest degree of efficiency
EAM technologies	Comprise (but are not limited to) computerized maintenance management systems, calibration management, tracking applications, the use of predictive maintenance software	A major enabler of EAM, using engineering data to provide support for MRO processes
Engineering data content	Electronic information that defines an organization's asset base, inventory stock, operations, resources, and maintenance	Fundamental to the execution of EAM strategy as its extracts value from the available data
People	People = employees	People are responsible for forming, executing, tracking, and managing all of the abovementioned components. Thanks to people, the EAM structure can maintain its integrity

assets. Management systematic approach gives also guidelines for industrial companies that would be managing customer's property as a business. Based on the work on enterprise asset management by MacArthur (n.d.), the following five building blocks of EAM are outlined: strategy for EAM development, descriptions of MRO (maintenance, repair, and overhead) processes, technologies that support EAM, engineering data content, and human resources to maintain the targeted service levels (Fig. 7.3). Table 7.3 presents a detailed overview of the elements of EAM and its functions.

Chapter 8
Role of Technology in Servitization

The role of technology is increasing in service businesses (West and Pascual 2015). Product information and product support are offered typically over the Internet. Connected products are smart and collect data from usage. This data can be transferred into centralized servers and big data type analytics can be performed. Additionally, service-dominant logic can serve as a basis for input-output analysis for measuring technology spillovers in service sectors (Hsieh and Yuan 2015).

8.1 Technology and Servitization

The use of advanced information and communication technology is certainly reshaping services and customer experience. The concept of customer contact becomes crucially important in the context of the application of modern technologies in service delivery.

According to Froehle and Roth (2004), the development of information and communication technology is influencing the interactions between customers and service providers, which may also influence customers' perceptions of services. Froehle and Roth (2004) outline four possible modes of customer contact in relation to technology. The modes are as follows:

- Technology-free customer contact (*face-to-face* service encounter; technology does not play a direct role in service provision)
- Technology-assisted customer contact (technology as an aid to improve *face-to-face* service encounter; the customer does not have access to technology)
- Technology-facilitated customer contact (*face-to-face* encounter where both customer and service representative have access to the same technology)
- Technology-mediated customer contact (where the customer and service representative are not physically connected and some form of technology is employed to enable communication; *face-to-screen* mode)

Fig. 8.1 Conceptual modes of customer contact in the context of technology. Adapted from Froehle and Roth (2004)

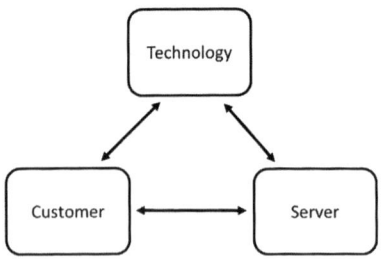

Table 8.1 Total service concept: elements and examples (Adapted from Fitzsimmons and Fitzsimmons 2004)

Elements	Description and examples
Core service	
Supporting facilities	Facility layout, access to service stations; in a wider sense, this may also cover web portals, mobile applications
Facilitating goods	Physical items related to service—spare parts, documents delivered
Facilitation information	Information items related to service—schedules, pricing schemes, product data, usage reports
Explicit services	Transactions completing the service: changing wearing parts in a machine, repairing control system
Implicit services	Intangible side of the service delivered: reliability of the system, feeling of being in control
Peripheral services	Services that are additional to the core service

- Technology-generated customer contact (face-to-face encounter is replaced by technology; *face-to-screen* mode)

Figure 8.1 presents an overview of the conceptual modes of customer contact in the technology context (Table 8.1).

8.2 Internet and Connected Products

Using technology to increase the servitization is led by companies in consumer businesses. Technologies such as the Internet of things (IoT), Industry 4.0 initiative, and big data analytics all give possibilities to transfer businesses. Potential implication strategies include the following:

- Physical products creating ecosystems. A software ecosystem such as Apple's app store is a well-known example of using the power of developers to increase the value of the product. Opening business possibilities to external companies is not a typical strategy. As products have software components, opening this part for external developers could work. Responsibilities and liabilities need to be fully agreed as many industrial products are expensive and may involve large risk.

- Transition from ownership to membership requires many changes in finance and operations. However, this should enable a stable revenue stream for the future.
- Open data from the product for analytics purposes could enable new uses and businesses. Opening the data from actual use of the system allows integration to related systems and new value added being generated by others.
- Applification—building mobile applications to connect the physical product and users is used in many product categories. The challenge is to develop such applications that they actually deliver some value to users and help to get closer to users by using a continuous copresence.

8.2.1 Architecture of Smart Physical Products

Many industrial products such as machinery are embedded with computers and smart by design. Software performs analytical and controlling functions based on sensor information as well as preferences set by the user. As mobile networks are pervasive in almost any location of the world, connected products are becoming standard. Connection may be established directly from the user to the machinery, or the machinery may contact a centralized cloud service which provides functionality for data analysis or even remote control. The main components in such a system are (Fig. 8.2):

1. Connected physical products collecting sensor information from the device and the environment, processing data, calculating daily averages, and logging events on a local machine. The system may detect its status and change the operational

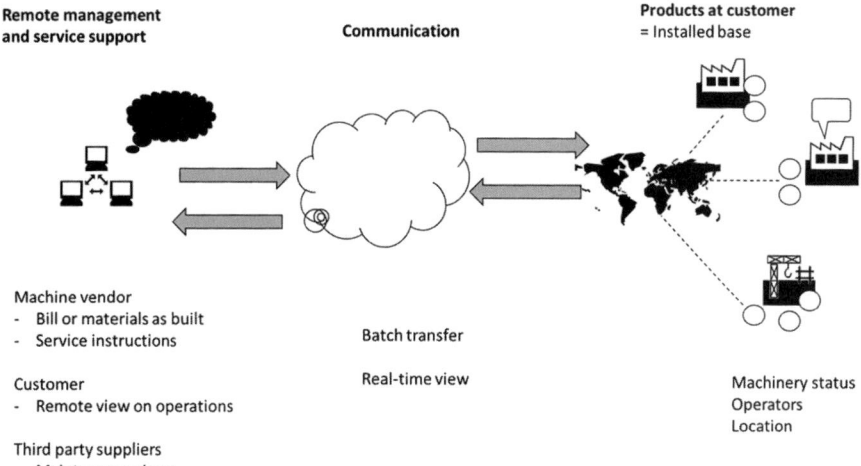

Fig. 8.2 Installed base connected to centralized cloud system

and communication profile accordingly. Local displays may provide a view of communicated items. Typically, data storage on the local machine level is limited to a few weeks or months, and data is aggregated into larger entities.

2. Communication infrastructure to send data periodically to a centralized cloud system; the system can offer synchronous and asynchronous data exchange, and in the case of varying data connectivity or a given communication budget, the system may exchange different sets of data. VPN may be used to enhance data security.

3. A centralized cloud infrastructure receives sent data from the installed base. Each machine is identified by using a unique ID and data may be grouped according to customers. Advanced analytics may be provided and data access can be given to customers, service personnel, and research and development. The actual value-added services such as mobile apps are built to use the data stored in the centralized cloud system.

8.2.2 Remote Management Systems in Service Products

Continuous remote presence on installations can enable new opportunities beyond diagnostics. Figure 8.3 illustrates examples of possible service products that a remote management system can create:

- On the management level, typical applications are related to support asset management over the life cycle of product use. Fleet management functionality may be offered to see the entire fleet at a certain customer or site location. A more detailed view of daily operations can be provided at the installation level. Process

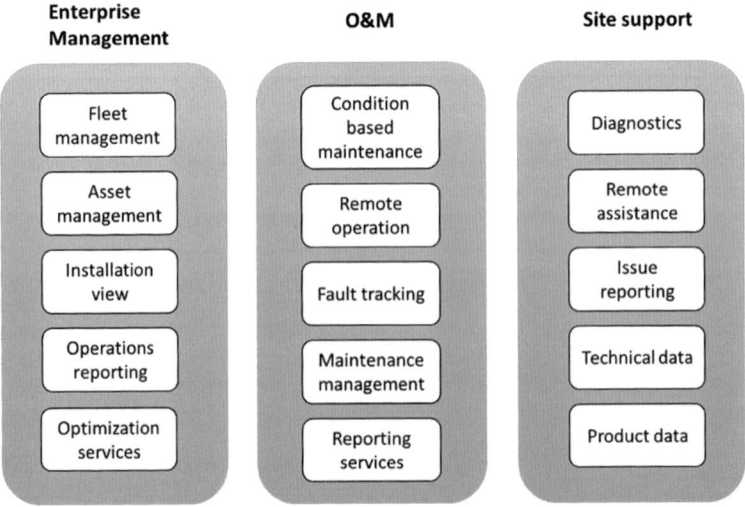

Fig. 8.3 Classifying applications built on remote management

8.2 Internet and Connected Products

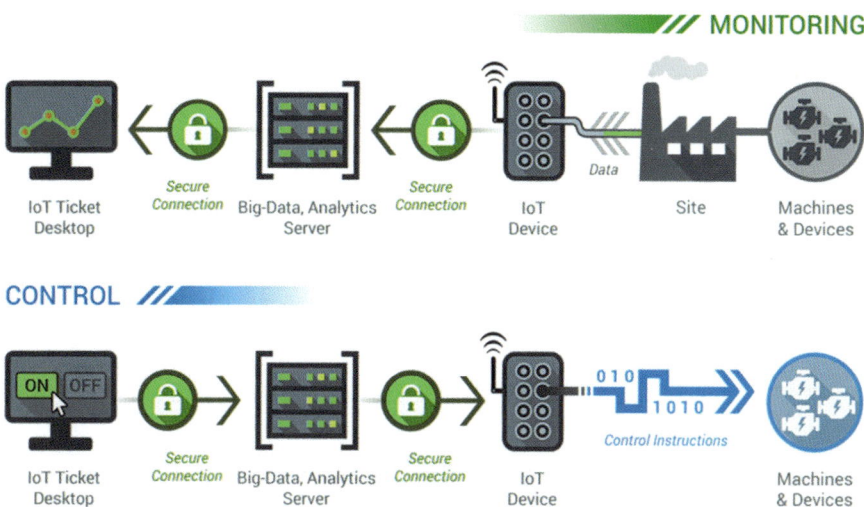

Fig. 8.4 Communication architecture of centralized data collection (Courtesy of Wapice IoT Ticket)

optimization and high-level decision support may be offered. Management reports of operational use and specific reports such as emissions can be generated as a service.
- Operations and maintenance (O&M) is the next level. Typical applications are related to condition-based maintenance, remote operations to support the work, fault tracking and monitoring, as well as daily maintenance management.
- Site support functionalities help the operators of the machine to access the information related to process and product. Diagnostics and remote assistance may be given by using the resources at a central service location. Local-level issue tracking may be kept along with logbooks. Up-to-date product information according to as-maintained bill of materials can be provided to users. This can help in the case of spare part orders and communication with the central.

From technological perspective utilization of Internet of things approach requires smart machines that are able to access local device data. The data can be sent periodically over the Internet to centralized cloud system which combines data from multiple machines. Secure VPN connections and authentications are needed in industrial applications. Centralized servers offer data storage functionality and tools to retrieve data for analytics. Bespoke user interfaces may be created to support each user group and access level which may be based on roles, organizational hierarchies, or temporary permissions (Fig. 8.4). User interfaces are typically responsive, but sometimes traditional PDF or paper-based reports can be generated for email distribution (Fig. 8.5).

Example Caterpillar's condition monitoring is based on analyzing data from a combination of condition monitoring components that vary from simple inspections

Fig. 8.5 Remote view of asset—power plant dashboard (Courtesy of Wapice IoT Ticket)

and regular fluid analysis to careful tracking of electronic data and analysis of equipment history. This helps to accurately assess the health and operating condition of critical equipment (Caterpillar Website 2014).

8.3 Industrial Standards for Managing Service Platforms

In addition to the business side, technology developments especially in information systems and telecommunications are driving the development of novel service products. Remote diagnostics and operations can be organized in the installed base of products. This side of the development is driven by large industrial companies including IBM, General Electric, and ABB.

Open O&M is an industrial forum for developing practices and standardizing operations and maintenance. It packages several existing technologies in the application area related to O&M and encourages the use of open interfaces between systems. The MIMOSA (Machinery Information Management Open Systems Alliance) group has developed "the adoption of open information standards in manufacturing, fleet, and facility environments—information standards which enable collaborative asset lifecycle management." There are two important frameworks, namely, OSA-EAI for enterprise application integration and OSA-CBM for condition-based maintenance. The figure below illustrates an overall architecture of how an open anticipatory logistics management system should be composed of reliability management and maintenance management elements supported by sensor data (Table 8.2).

Another technical standard which guides how to build a condition-based management system from several layers and building blocks is ISO 13374. This standard layers functionalities in different levels, and part of the operation can happen close to the actual device (machine), certain things on the system level (site), and higher-level strategic decisions on the fleet level. Figure 8.6 shows the bottom-up process steps of condition-based maintenance:

1. The data acquisition layer is responsible for collecting data from sensors and records according to a given sampling interval.
2. The data manipulation layer provides simple calculations to modify the data into correct units. Daily minimum, maximum, and average values may be recorded on local storage.
3. The state detection layer assesses the exceeding of any given warning limits. In case a sensor value exceeds the triggering point for a certain time period, the system state change is recorded.
4. The health assessment layer combines sensor data with product information and can diagnose potential causes of the issue.
5. The prognosis assessment layer makes an actual forecast for the system or component—what is going to happen in the foreseeable future based on previous experience and statistical data.
6. The advisory generation layer is the highest level. It is a decision support system that actually generates possible actions for the operators. Advice is given to local operators and maintenance personnel and system behavior can be monitored remotely.

There are business opportunities for service development when real-time information or close to real-time information retrieved from devices and delivered to the manufacturer can enable many applications that can be part of service product offerings. Some examples include documentation for a certain part (e.g., based on engine serial number) being connected to e-business systems for spare part ordering. Integrating condition-based maintenance into the scheduling of maintenance personnel and spare part logistics may give new perspectives for the customer.

Table 8.2 Possible service products enabled by remote management

Service			Online support
Service products	Product information sharing services	Cost-benefit calculations	Tracking savings from actual use
		Product configuration for sales	Product configuration parameters as initially planned
		Documentation online	User manuals
			As-maintained bill of materials
		Customer support by phone and online	Phone, email, video
	Operations optimization services	Installation service	Product registration and support
		Delivery service	Tracking deliveries
		Technical support for similar products	–
		Repair service	Diagnostics online
		Maintenance	Data records from operations
			Logbooks
		Spare parts	Electronic ordering
		Recycling service	–
		Product upgrading	Software-based product upgrades
		Extended warranty	Monitoring warranty-related conditions
	Business services	Operating customer's product	Remote control
		Operating customers process	Remote control
		Procurement service	Electronic ordering
		Warehousing products for customer	Electronic ordering
		Key performance indicator dashboards	Remote management for fleet/assets
		Reselling customers used machinery	–

8.4 Condition-Based Maintenance

According to Ellis (2008), condition-based maintenance is a management philosophy that postulates repair or replacement decisions on the current or future condition of assets. Therefore, the main reason for executing maintenance is a change in condition and/or performance of an asset, which implies that the optimal time for performing maintenance can be determined by regular monitoring of an asset.

8.4 Condition-Based Maintenance

Functions from ISO 13374: Machine Condition Assessment Data Processing & Information Flow Blocks

Level	Block	Example
Centralized server	Advisory generation	Suggest replacing the component in two weeks from now
	Prognostics assessment	The component will last 200 hours
	Health assessment	Diagnose: Component Y766 has problem.
	State detection	Rule triggered. Detect exception and send info to portal
Local machine level	Data manipulation	Calculate daily min, max and averages
	Sensor	Record temp, amps and volts 1/s

Fig. 8.6 Condition-based maintenance as defined in open O&M and ISO 13774

The main objective of condition-based maintenance (CBM) is to minimize the cost of inspections and repairs by systematically collecting, analyzing, and interpreting data. Therefore, successful CBM requires understanding of failure modes and rates, asset criticality, and potential payoffs associated with different maintenance strategies.

Ellis (2008), based on a comprehensive literature review, outlines the following benefits of CBM as:

- Greater ability to maximize the service life of a component (as compared to, e.g., time-based preventative maintenance)
- Ability to take preventative action just in time while avoiding the serious consequences of a machine breakdown

Coetzee (1999) proposes a holistic approach to CBM. The author suggests that the holistic approach should include evaluation of maintenance assets. According to Shohet (2003), effective implementation of CBM requires the development of performance indicators for building components and systems. Ellis (2008) also claims that the effective application of CBM also requires the use of analytical tools such as failure mode and effect and criticality analysis (FMECA) to determine the likelihood of a failure and how this failure would occur. Moreover, a reliable information tool for capturing and tracking repairs and their associated costs would also be needed.

There also are several possible disadvantages to CBM, which are as follows:

- Installing condition monitoring test equipment is expensive as is also analysis of the available data.

- Training staff also incurs costs and it is necessary to have knowledgeable professionals to analyze the data.
- Fatigue in uniform wear failures is not easily detected with CBM measurements.
- Condition sensors may not survive in the operating environment.
- Unpredictable maintenance periods (Maintenance Assistant Webpage 2014).

CBM often occurs in the phases presented in Fig. 8.7.

There are several types of CBM implementations and Table 8.3 presents a summary of the most important ones. The sensor and analysis are based on the type of component that defects or wears out.

Condition-Based Maintenance at Wärtsilä

Wärtsilä is a Finland-based manufacturer of diesel engines and marine power systems. The company has almost 50 % market share in medium-speed marine engines and 18 % in larger low-speed main engines. The company has service contracts beyond its own products. Also other brand name products are maintained.

CBM in Wärtsilä is aimed at optimizing the availability, reliability, and performance of installed equipment. CBM is a part of a wider-reaching concept of Dynamic Maintenance Planning (DMP), which consists of the following:

- Remote condition monitoring service
- Site audits and intermediate/opening inspections
- Maintenance planning services

Having probably the world's largest customer base in this domain with regard to installed base and service contracts presents a great source of data analysis. The company has provided a condition-based maintenance service which has been built on online diagnostics and remote advisory. The system is based on onboard computers collecting data from each part of the ship automation system: engines and propulsion systems producing large quantities of data from each sensor. A ship automation system may consist of tens of thousands of values monitored in a millisecond level.

Data is stored locally, and summary is sent over to Wärtsilä CBM service engineers by using satellite connections. Standard analytics are collected from the sent data, but ad hoc queries can be processed and large data dumps can be analyzed above the single ship level. The company has systematically developed the system to include new features allowing customers to access the fleet data by using a portal system.

Data mash-up is also one of the key concepts. Location and ambient environment data including weather forecasts can be combined into analysis as well as the maintenance history of each vessel. This enables combining functionalities from fleet management and asset management to condition-based management and maintenance planning. For example, maintenance can be scheduled by combining condition analysis with route and weather information as well as spare part and maintenance crew availability information. Another important feature could be the optimization of fuel consumption by providing analyses taking into account crew, weather, and operation profile of the ship. Having access to a great number of seagoing vessels' operational data can provide important benchmarking information for ship lines.

8.4 Condition-Based Maintenance

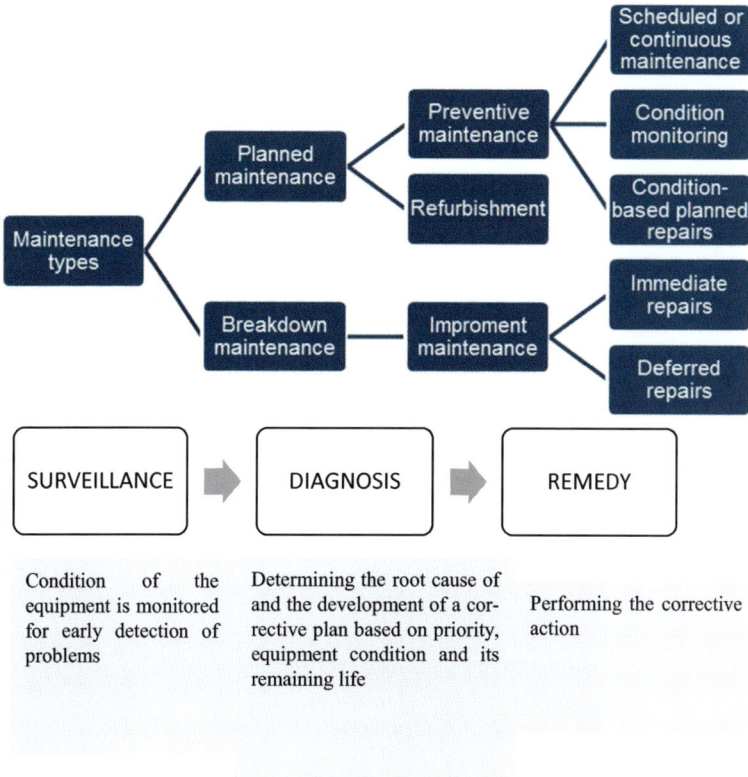

Fig. 8.7 Phases of CBM. Adapted from Maintenance Assistant Webpage (2014). ISO 13306 maintenance types

Table 8.3 Types of CBM

CBM type	Specification
Vibration analysis	Certain rotating equipment exhibits a degree of variation which, when it becomes excessive, can be monitored by sensors
Infrared	Using IR cameras to detect high-temperature conditions in energized equipment
Ultrasonic	Used for the detecting of deep surface defects such as corrosion
Acoustic	Used for detecting gas, liquid, or vacuum leaks
Oil analysis	Measuring the number and size of particles to determine asset wear
Electrical	Motor current readings using clamp on ammeters
Operational performance	Using sensors throughout a system to measure, e.g., pressure, temperature, flow

Source: Adapted from Maintenance Assistant Webpage (2014)

From the technology point of view, the Wärtsilä case presents a dispersed data storage system collecting large quantities of data from local automation systems and collecting real-time views into a centralized system for analysis and decision-making. The data storage is a combination of various techniques, and different views are available for each user according to profile and authorization.

The business case is based on creating a new type of value added for customers. The concrete examples are new products which use real-time usage data, mash-up it with other data streams, and enable measureable benefits such as fuel consumption optimization or reduced maintenance. The other obvious value added for Wärtsilä internal operations is to have a better visibility toward end users—how products are actually used and how they perform under different conditions. This also supports strategic transformation of the company to change its position in the value chain from component supplier to a key solution provider with close proximity to end customers.

In the case of Wärtsilä, the CBM is organized around the following principles:

The operational data on assets (in this case—engines) is collected during normal operations and evaluated by experienced staff with the help of appropriate evaluation tools.

- Data collection is done automatically—no additional work for the personnel is generated.
- The results from the evaluation are reported on a monthly basis with operation data, notifications, and advice.
- Remote condition monitoring lies at the core of Wärtsilä's CBM system.

Timely and accurate maintenance offers benefits such as improved reliability, with unexpected downtime eliminated, as well as improved availability and efficiency of assets. Remote condition monitoring is one of the core elements of an efficient CBM. Well-performed remote condition monitoring results in detailed reports and real-time data available on the server. The available data consists of, e.g., reports with recommendations for corrective action and preventative maintenance. The data is often available in real time, is stored on servers, and can be accessed after logging in, and more importantly, this is all achieved without physical contact with a device (Condition Based Maintenance in Wärtsilä 2014).

8.5 Cloud-Based Services and Portals

Online services and portals are offering a novel approach to connect with end users of the machinery. Typically, the services are offered as web-based portals that require registration of users and machinery they use:

1. User manuals—up-to-date user manuals for individual machinery. Complex products tend to have large documentation, which typically is delivered on installation and archived. The information content is not accessible by the operators and maintenance personnel on paper copies. Online services solve this problem and ensure better utilization.

2. Service bulletins are updated and revised instructions for operators and maintenance personnel. This information sent by the machine vendors to actual users of the machinery is typically time critical, and it is important to ensure that the updated instructions are safely received.
3. Spare part web shop—spare parts and consumables may be offered by using a web shop. Technical drawings may be combined with the part catalogue to ensure that correct parts are ordered. Tracking and tracking orders can be integrated to online shop functionality.
4. Maintenance log—an online logbook for periodic maintenance and service events.
5. Help desk and support—ticketing systems are used to manage the flow of support request and technical advisory needed by the customers. Key performance indicators for service delivery, including waiting time, time to solve case, and customer feedback, can be integrated. Volume numbers of tickets at each stage in the pipeline are typically used for capacity management of the servers.
6. Diagnostics—diagnostics may be based on user entry and flow charts to find out resolutions for operational or maintenance-related problems. Another alternative is to upload error log from machinery and process the analytics part automatically or assisted by a team of experts. This type of features is getting more common as device sensors record a lot of data, and other ways of transferring information—emails and phone conversations—are not very good in sending large quantities of complex data.
7. After-sale connections from marketing perspective may be very powerful as users and the machinery they use are well known. Additional services, add-on products, and spare parts may be offered in combination with online shops.
8. Updates—software updates and additional functionality can be updated from the centralized cloud. This makes it possible to maintain system remotely and can justify monthly service fee of the product as it is very common in the enterprise software sector.
9. Fleet management approach combines several machineries of the same or different type into a grouping level. Key performance indicators may be used to monitor equipment overall efficiency, and based on this the users may drill down on individual machinery to see the details. Fleet management functionality is typically focusing on utilization of assets.
10. Service-level agreements—managing contracts and service-level agreements is a common feature of online portals. End users can see the status of the contract and upcoming checkpoints of performance evaluation.

8.6 Big Data Analytics

Big data analytics refers to the use of information technology and statistical tools to manage large quantities of data generated in high speed, and it may include large variety of data types. For industrial service business, the data sources may include customer order data, data generated by the machinery, installed base, production data such as bill of materials and test reports, service orders, and auxiliary data such as weather and maps.

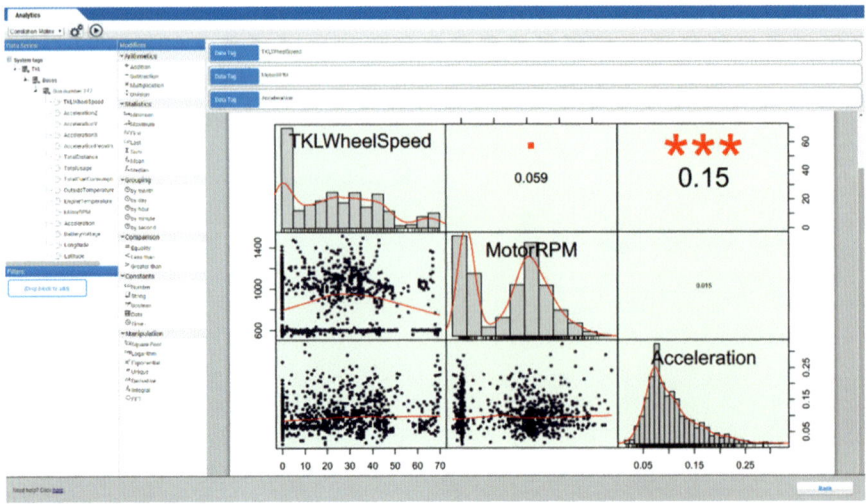

Fig. 8.8 Analytics display of commuter bus—correlation matrix visualized

Operational data collected from the assets can provide an interesting view for data analytics. Typical use cases for service development include the following:

- Analytics for operational reporting
- Analytics data for ad hoc R&D queries to support goods-based product development
- Supervisory monitoring and control of the asset utilization and operations
- Operational efficiency and KPI tracking for asset and fleets of assets
- Condition monitoring and maintenance
- Regulatory reporting

Figure 8.8 illustrates an example of commuter bus data: relationships of speed of the vehicle, motor RPM, and acceleration analyzed by using a correlation tool.

8.7 Applification

Mobile applications connecting the user, the asset, and the vendor are becoming increasingly popular among machine builders. As users are carrying Internet-connected tablets and mobile phones, providing an access to machinery is feasible. Mobile apps can support several daily operations for operators, maintenance staff, and managers overseeing processes. Mobile apps are typically not connecting to asset directly but a centralized cloud server which maintains linkage to products. Specific features of industrial service mobile apps include:

- Authentication of the user and providing this information to the system.
- Location-based services are valuable for mobile machinery.

8.7 Applification

- Communication linkage to support and advisory services.
- Opening support cases by linking machinery data, user requesting, and ambient environment information.

The concept of "applification" refers to increasing phase of introduction of professional mobile applications running on mobile phones or tablet computers. Mobile applications are offered to connect the user with the vendors of the machinery and sometimes directly with the machinery. Certain key characteristics are typical for mobile apps offered:

1. Portals and general connection to vendors—Contact information for after-sales and technical support is typically offered as well as basic user interfaces for user instructions or diagnostics. Customer portals may be connected to mobile application to show the status of services or ongoing processes within vendor interaction.
2. Location-based services are utilizing the GPS functionality of the mobile device. Based on location, the closest contact points may be shown on the map, and in the case of mobile machinery, the exact location of equipment may be presented to users.
3. Remote monitoring functionality requires that machinery is connected to the Internet. This type of features of software retrieves operational or maintenance-related data from the machinery and shows to mobile application user.
4. Limited remote control—Mobile applications may allow limited remote control, which is not safety critical. Typical uses include setting up machinery parameters, remote warm-up, or shutdown.

For the machine vendors, this type of approach offers a view on operators and end users. Assets and actual users can be linked with each other. From service business point of view, the potential benefits for the machine vendor include:

1. Accessing end users and communication between the operators and the machine vendor
2. Providing access on operational data from machinery
3. Streamlining service processes such as spare part delivery process or fault diagnostics

Chapter 9
Pricing Decisions: From Ownership to Subscription

The Trendwatching (2011) report emphasizes that *traditional ownership* is often a burden for modern and usually mobile customers. Therefore, *fractional ownership* that partially relieves customers from responsibility, costs, and commitment is gaining popularity. Such a form of ownership is particularly interesting for consumers who value experience over possessions. *Planned spontaneity* allows customers to use bulky and otherwise rarely used items. With online access, customers are able to book them in advance and use when needed.

9.1 Subscription Services

Over the years, there has been a shift from ownership, through fractional ownership, toward ownerless, lifestyle leasing business models, and planned spontaneity. Many organizations are constructing innovative business models that are far away from the traditional notion of ownership. These attempts are generally targeting consumers who are willing to adapt to new possibilities and typically developing the service offering by active participation. Consumer services are driving this transition and industrial services are following. Table 9.1 presents a short overview of the innovative approach to ownership. Recent successful examples have been related to the media industry as intangible information delivery costs are minimal and the offering can be expanded with minimal costs. In this field well-known subscription examples include on-demand television, e.g., Netflix, music (Spotify), books (Kindle Unlimited), and ScienceDirect in academic journals provided to research institutions. Now shared ownership and subscription have moved to the tangible goods side as well, and subscription-based cars, food, and clothes have been introduced to markets.

Payment models are related to ownership of the product. The spectrum between owning a pure physical product and utility service based on subscription and actual use is very wide. Some type of ownership and control is required in central assets;

Table 9.1 Innovative approaches to ownership

Company	Innovative offering
WhipCar	An alternative to car rental and ownership. Renting cars from private car owners (collaborative consumption, peer-to-peer rental)
Girl Meets Dress	Dress and accessories rental for any occasion. Easy return after the event
The Larder Box	Monthly subscription to a service that delivers food and recipes to the customer's door
Boris Bikes	Subscription for bike service – ride when you like and then return the bike for the next customer to use
FlexPetz	Shared pet ownership (annual, monthly, or per-visit charge)
Netflix	Subscription based, access to movies and television series for a monthly fee
Spotify	Subscription based, access to music for a monthly fee
Kindle Unlimited	Subscription-based unlimited supply for Amazon e-books
Birchbox	Subscription based, monthly deliveries of cosmetics
Peach Pass	Front-of-the-line subscription model. Subscribers are allowed to use fast lanes which enable them to avoid traffic: operates in Georgia, USA

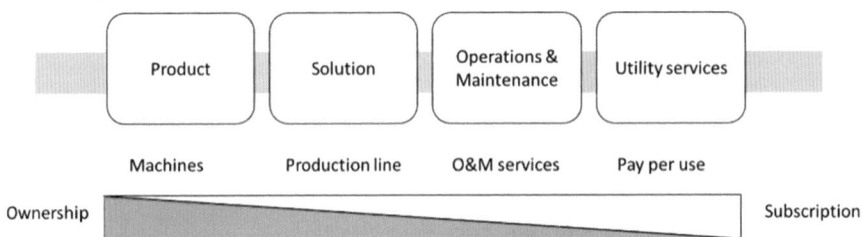

Fig. 9.1 Ownership-subscription spectrum

less focal products may be leased. However, there are several shades in the spectrum, and most of the products have both a physical component and at least some kind of service element (Fig. 9.1). Utility types of services provide an opportunity for flexible operations and less commitment to physical ones. There is a trade-off – increased flexibility also means less control. Power has been given to the service provider from the customer. An important assessment needs to be made in each case – is this asset for the user or something less important to be dealt with as a utility?

One approach to analyze the possibilities in pricing and charging logic is to consider the example of transportation. Figure 9.2 illustrates different forms of ownership and payment method in the specific case of traveling by car. On the two separate ends is either owning a car or using a taxi. However, a customer has other types of services that come in between, such as partly owning a car, leasing, or renting it. As an alternative solution, a customer can choose to travel by bus which implies paying only for the ticket (per use).

9.2 CAPEX and OPEX

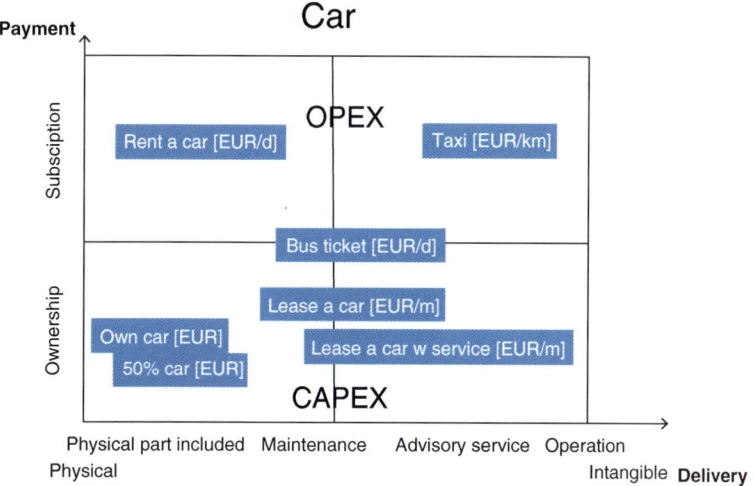

Fig. 9.2 Different forms of ownership

9.2 CAPEX and OPEX

In an industrial business to business environment, the justification for subscribed services needs to be supported by financial results. Capital and operational expenditures are a part of a larger total ownership cost (TOC) concept which is aimed at analyzing and understanding the true cost of doing business with a supplier, and the analysis goes beyond the price. Therefore, TOC can be defined as both a tool for purchasing and a philosophy (Ellram 1993).

In the simplest terms, capital expenditures (CAPEX) are usually related to major investments in goods, which are then shown on the balance sheet and depreciated over the life cycle of the product or an asset. Operating expenditure (OPEX) is shown on the profit and loss statement. In other words, these expenses are considered to be incurred continuously.

Decisions on selecting CAPEX or OPEX should be taken by a company and based on a thorough analysis of the capital expenditure structure. CAPEX and OPEX decisions (such as moving to a pay-as-you-go model) mean that a company's cash flow is changing. Frequently, organizations prefer to lease rather than purchase, which stems from being limited by the markets or private lenders in the amount of capital expenditure they are able to access. For this reason, many companies would wish to direct their investments toward activities generating sales, which translates into preferring to lease rather than purchase (Logicalis Australia 2015) (Table 9.2).

Table 9.2 OPEX and CAPEX comparison

OPEX (operational expenditure)	CAPEX (capital expenditure)
Business expenses incurred in the course of ordinary business, such as sales, general and administrative expenses	Business expense incurred to create future benefit. In other words, it is an acquisition of assets that will have a useful life beyond the tax year
Example: buying a laser printer	Example: buying paper and toner for printer
Paid as a lump sum (or financed, with extra charges)	Paid monthly

Adapted from diffen.com (2014)

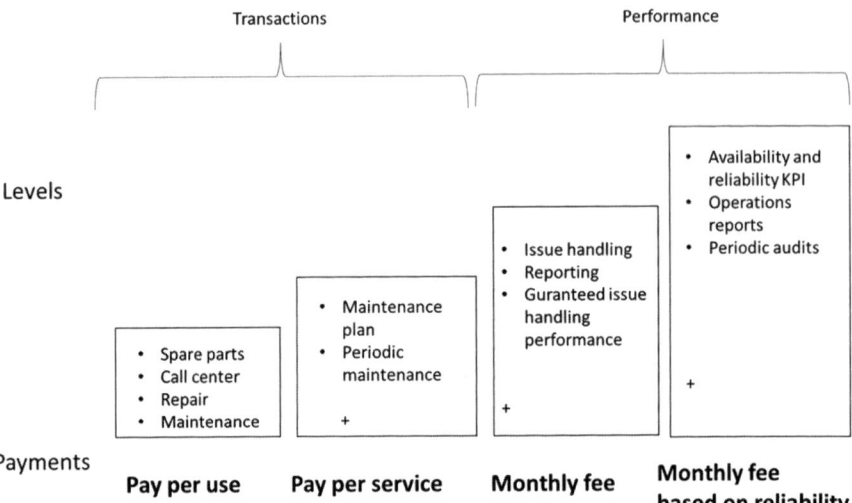

Fig. 9.3 Different forms of ownership

The transition from CAPEX focused thinking toward OPEX is often supported by stretching the product offering toward performance and life cycle-based service products. In the first phase, these do now need to be completely replacing the traditional offering but seen as a comparable alternative in terms of indicators. Figure 9.3 shows four steps in service offering development. First one from the left, pay per use is sales of spare parts and repair work; pay per service is planned maintenance which includes the works and parts required by the machine-specific plan. Warranty may be tied to this level. The next level is monthly fee which includes the previous level packages as well as support for operations; performance guarantees may be given as well as periodic reporting of operations efficiency. The ultimate performance level system is monthly fee based on system reliability. In this level, performance bonus or penalties may be given based on periodic performance reviews.

9.3 The Subscription Economy

Buying a product or services based on a predefined, e.g., monthly, rate is not particularly novel since companies such as Netflix or Spotify have long built their successful business models based on subscription. Nevertheless, literature on the topic is scarce.

The subscription economy is one of the main emerging trends, and subscription-based business models are one if its manifestations. Currently, various types of services and products are being offered for a flat-rate monthly fee. Successful examples of companies such as Netflix or Spotify have sparked a discussion on whether switching to a subscription-based business model should occur on a wider scale. In the current world, it has become evident that consumers value the possibility of using products or services without being forced to own them; on the other hand, not every switch to a subscription business model will always be successful, as it is risky to assume that organizations will easily repeat the success of the abovementioned Netflix or Spotify (Fortune 2014).

Subscription-based business models offer benefits to customers and entrepreneurs, and one of the major benefits is the possibility of smoothing variations in demand. While customers are getting the same orders on a fixed date, manufacturers or service providers do not need to worry about not being able to fulfill the needs of their customers on time. Moreover, a shift toward a subscription-based business model can mean that an organization increases its chances of surviving the recession. By far the greatest benefit that a subscription-based business model can bring to an organization is increasing its own valuation, as recurring revenue is considered to considerably boost the value of a company for potential buyers (Inc. 2014).

In summary, subscription-based business models still have great development potential and their true impact still remains to be seen.

9.4 Pricing Models

According to Kindström and Kowalkowski (2014), pricing capability is an essential element of generating revenue models. The authors emphasize that pricing capability is required to define the charging mechanism for new service products. Sometimes existing products and services need changes in revenue generating models. The authors also claim that value visualization capability is essential, especially when organizations begin with offering their services for free. In such a case, introducing service fees requires keeping customers aware of the added value they are paying for.

Rehme (n.d.) outlines the following most common pricing models in services:

- Cost-plus pricing
- Target-return pricing
- Competitive pricing
- Life cycle pricing

Table 9.3 Most common pricing models in services (and products)

Pricing model	Description
Cost-plus pricing	Adding a standard markup to the cost of a product requires certain information about costs, ignores current demand and competition, and supports inefficiency.
Target-return pricing	Aimed at achieving a target return on investment over a specified period of time.
Competitive pricing	An organization can price its service (or product) based on the current market structure – at market price, above or below.
Life cycle pricing	Market penetration pricing (for highly sensitive markets where low prices stimulate its growth).
	Market skimming pricing (for markets with a sufficient number of buyers with high current demand and relatively low unit costs for producing smaller amounts).
Experience curve pricing	Based on the dynamics of cost, organizations adjust their prices accordingly, and buyers and sellers adjust to one another as time goes by.
Value-based pricing	Based on the value which the customer derives from the use of service or (product), this strategy is especially useful when introducing an improved product or competing against a well-established rival.

Adapted from Zeithaml et al. (2006)

Table 9.4 Pricing models of managed service providers (MSPs)

MSP pricing model	Description
Monitoring only	MSP provides a service of a network that monitors and alerts.
Per device	Each device supported in a customer environment is a subject to a flat fee.
Per user	A flat fee for support of all devices end user is using is invoiced per end user on a monthly basis.
Tiered	Several bundled packages of services with each increasingly more expensive package providing more services to customers.
All you can eat	A very flexible pricing model that includes all remote support, on-site support, as well as lab or bench time for an entire organization for a flat fee per month.
SLA-based pricing	Pricing strategy based on the risk to MSP and what kind of risk is taken on behalf of the customer.
Value-based pricing	Based on understanding the needs of existing or prospective customers and setting prices on the perceived value of services.

Adapted from Techtagert Website (2014)

- Experience curve pricing
- Value-based pricing

Table 9.3 provides a brief overview of the models in detail.

Table 9.4 presents an overview of the most common pricing models outlined on the basis of managed service providers (MSPs).

Joy Global, a US-based company manufacturing machinery for mines, is a good industrial example of changing the price scheme from capital expenses to operations.

Case Example: Joy Global Mining ("Smart Services" and "Lowest Cost Per Ton to the Customer") Joy Global is one of the leading companies in the development, manufacturing, distribution, and service of underground mining machinery for the extraction of coal and other bedded materials. The company offers smart services that exceed the idea of traditional product offering and associated services. Joy Global aims at gathering relevant information and using it to deliver smart services, which are directed at providing the customers with better machine availability, utilization, as well as minimal downtime and reduced costs. The company also employs life cycle management (LCM) that is aimed at providing its customers with the lowest cost per ton over the life of the equipment. For certain product systems, the company offers subscription-type contracts. The company has a customer promise for the "lowest cost per ton to the customer." In order to deliver the service, the charging logic is based per ton contracts. In practice, the availability and operational performance of machinery are measured by using remote monitoring systems, and a bonus/penalty structure can be used (Joy Global Website 2014).

9.5 Freemium Pricing Model

The term freemium pricing model was coined by combining the words "free," meaning the core product that is available free of charge to a large group of users, and "premium," meaning selling premium products to a smaller fraction of this user base.

Freemium originates from the software industry and has successfully diffused into newspapers, music, publishing, telco, education, and many more. The concept can be successfully developed thanks to the fact that today it is possible to create and distribute various value propositions by using ICT. Once software has been developed, the distribution cost is practically zero unlike in goods-based products. Freemium is rooted in design thinking, where, in simple terms, ideas are generated as new and bold rather than stemming from the careful consideration of alternatives.

Freemium does not imply a free trial, as it does not include any trial period. The core product is always free, and users can always move to adding premium features that offer and improve the experience.

Freemium-based business models are currently the most popular among a wide range of industries. For example, apps with freemium models account for 98 % of the revenue in Google's app store and 95 % in Apple's app store (www.freemium.org, 2015) (Table 9.5).

9.6 Software Ecosystem Model

According to Bosch (2009), software ecosystems are logical development steps for enterprises that have already created a successful platform and an intra-organizational product line. Transition to a software ecosystem is commonly achieved through

Table 9.5 An overview of premium cases as listed on www.freemium.org

Category	Case	Brief description
Application and games	Evernote	Full features of the version are free ("free extra brain"); premium users get other features tailored to specific needs.
Music and videos	Pandora	A radio service that plays music according to the user's individual taste. The free version is ad supported, while the premium version is ad-free.
Online services	LinkedIn	The free version offers a variety of opportunities to connect with like-minded professionals. The premium version offers extended features.
Online/offline	Starbucks	The company aims to prove that a coffeehouse can be much more than offering coffee but also various services. The free version comprises of:
		Free Wi-Fi Internet access (no restrictions)
		Free access to Starbucks Digital Network premium content such as *Wall Street Journal*, iTunes, the *New York Times*, etc.
		The premium version comprises of:
		Full-access subscriptions to these content providers
		Coffee

selecting a software ecosystem type, which, most commonly, is either application-centric software or an end-user programming ecosystem. Both imply transforming a product line into a platform of external developers, which requires decisions on the following issues:

1. Directed or undirected approach to partner selection and choice of application
2. Tiers of developers (as intermediate stages between directed and undirected approaches)

9.6.1 Directed Approach

This approach assumes that there are certain areas of functionality that have been identified as important/critical, but the company is either unable or unwilling to develop them itself. This calls for both selecting partners that are able to provide the required solutions and prepare the agreements. Usually, the agreement clarifies the issue of sharing revenues. Additionally, the partner company typically has included an access to the technical infrastructure platform as well as the products developed internally on top of that platform. Such a configuration requires special attention to intellectual property issues.

9.6.2 Undirected Approach

Contrary to the directed approach, in the case of undirected approach, a platform company offers the external developers an opportunity to develop applications on a platform basis. Moreover, there are no constraints in terms of who gets to develop applications or what kind of application is built. Based on an assumption that competition is the best mechanism to deliver the best solutions to both users and customers (who make the final decision on which selection of solutions to use in their daily operations), the external developers are free to build solutions that compete with each other as well as with the platform company itself (Bosch 2009).

Directed and undirected approaches are the two extreme approaches; however, approaches based on different tiers of developers can be implemented in order to mitigate the deficiencies of directed and undirected approaches.

9.6.3 Tiers of Developers

The possibility of having different tiers of developers moderates the deficiencies of both directed and undirected approach. Bosch (2009) outlines the following tiers of developers for a product line transforming into software ecosystem:

1. Internal developers – in the case of internal developers, the level of trust is higher and communication is easier; therefore, deeper access and insights to the platform can be offered.
2. Strategic developers – developers carefully selected by a company, with whom long-term relations have been built; they have a deep access to the platform, understand both platform and product lines, and often have access to private user data.
3. Undirected developers – these developers develop solutions independently with the aim of selling them to platform users. For security reasons these developers have restricted access to the platform and product functionality.
4. Independent solution vendors – these offer customer-specific integrations of one of more products provided by the product line company.

Chapter 10
Value Chain Effects

Value chains define the position of each actor of the supply chain based on how much value added each partner of the supply chain is delivering.

Porter (2001) defined value chain as a representation of a firm's activities that are performed to design, produce, market, deliver, and support its product. According to Porter (2001), value chain describes all activities that a company is performing and how they interact, which, in turn, is necessary to outline the sources of competitive advantage. Value chain segments a firm into its strategically relevant activities, which facilitates understanding the structure of costs as well as identification of potential sources of differentiation. Porter (2001) also emphasizes that the value chains of firms are embedded in a larger structure called value system.

According to Paiola et al. (2012), reorientation of a company's value proposition, which also reshapes the company's value chain, is caused by the commoditization of products, declining profitability, as well as the growing number of customers with increasingly complicated requirements. Therefore, manufacturers of industrial machinery are reinventing their value propositions guided by the principle of shifting focus from selling products to providing solutions. The authors quote examples of multinational companies such as Alstom, General Electric, or IBM.

10.1 Cocreation and Coproduction of Value

According to Lusch et al. (2007), cocreation of value denies the main assumption of goods-dominant logic, which implies that value is something added only to products in the manufacturing process. Service-dominant logic assumes that value can only be determined by the customer in the process of "consuming a service." The authors claim that the notion of value is also related to the concept of customer experience. Lusch et al. (2007) claim that one of the opportunities for companies to compete through service is the identification of innovative ways of cocreating value.

The authors also highlight the most important distinction between service-dominant logic and goods-dominant logic: namely, the importance of interactivity and doing things with the customer (as in service-dominant logic), as opposed to doing things for the customer (as in goods-dominant logic).

Identification of innovative ways of cocreation of value is a great opportunity for organizations to compete through services. According to Lusch et al. (2007), cocreation of value is largely dependent upon other entities of value creation such as resources. The authors claim that resources that are endogenous to value creation are often described in terms of belonging to a usually uncontrollable external environment. Lusch et al. (2007) also claim that "the customer is a primary integrator of resources in the creation of value through service experiences that are interwoven with life experiences to enhance quality of life" (p 12).

According to Vargo et al. (2008), cocreation of value is not limited solely to exchange within dyads of service systems. Value cocreation rather occurs *through the integration of existing resources with those available from a variety of service systems that can contribute to system well-being as determined by the system's environmental context* (p. 150).

Furthermore, the whole concept of servitization and offering complete product-service systems and creating value for customers as well as customers' active participation in value creation can be approached from a perspective of more effective utilization of resources and circularity (Tukker 2013; Lindahl et al. 2014). Therefore, it can be stated that the value chain effects reach far beyond the particular interests of customers and providers.

10.2 Supply Chains and Networks

Supply chains are systems and organizations involved in moving products and services from upstream suppliers to manufacturers, wholesalers, retailers, and finally end customers. The supply chain concept has evolved from analyzing operations and material management from the single business unit level to cover larger entities covering multiple business entities: the flow of goods, information, and finance in the chain. Supply chains are typically parts of larger, complex, and changing supply network systems, where interactions between business entities are more complex and dense (Fig. 10.1).

10.3 Vertical and Horizontal Integration

In order to be able to truly benefit from servitization, companies should be less concerned solely with providing services, but rather aim at building their competitive advantage on the basis of combining services with products in order to create high-value integrated solutions. The creation of these is usually achieved by starting

10.3 Vertical and Horizontal Integration

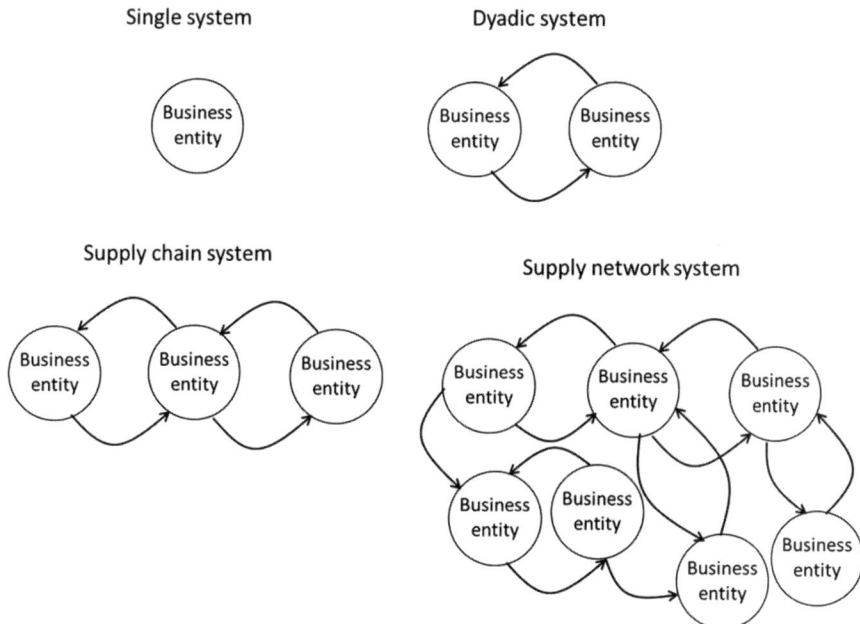

Fig. 10.1 Configurations in delivery. Adapted from Harland (1996)

from the existing manufacturing base and moving toward downstream into the provision of services to distribute, operate, maintain, and finance the product during its complete life cycle. The diversification of a firm's activities can be achieved through horizontal or vertical integration (Davies 2004).

Davies (2004) defines vertical integration in the context of the combination of successive stages in the flow of productive activities from raw material to a final product or service. Vertical integration can be achieved both backward and forward. The author defines horizontal integration as the combination of firms in the same stage of production.

Baines et al. (2011) define vertical integration in the context of traditional manufacturing as a situation when *a firm owns and takes responsibility for its upstream suppliers and downstream customers* (p. 984). Furthermore, vertical integration implies a company's engagement in different aspects of production. Backward vertical integration usually means taking over the activities of suppliers of inbound materials, while forward integration implies taking control of the outbound supply which has been managed by customers earlier (Baines et al. 2011).

In the servitization context, vertical integration can be observed in the context of product-centric services which require the provider to take on maintenance activities, earlier performed by the customer (Schmenner 2009).

Baines et al. (2011) also refer to the reasons underpinning vertical integration in the servitization context. The authors claim that companies are under pressure to fulfill contractual obligations to their customers combined with an internal pressure on delivering these as economically as possible. Furthermore, in the case of advanced services, customer value is centered on the outcome of services rather than on the service activities. Outcomes are measured through metrics such as asset performance, availability, and reliability. Service providers are therefore challenged to deliver them against the abovementioned metrics as economically as possible. The most important measure is connected with the cost of delivering an advanced service contract, which should be kept low for competitiveness (Baines et al. 2011).

Both vertical and horizontal integration will result in the need to develop certain capabilities, and companies will need to decide whether to develop them in-house or outsource (Davies 2004; Oliva and Kallenberg 2003; Paiola et al. 2012).

10.4 Moving Downstream in a Value Chain

Value chains have been studied in several industries in the context of how certain companies have secured their position in the value chain and are able to make a profit, while others may be succeeding much less (see He et al. 2010; Payne et al. 2009; Raddats and Burton 2011; Ritter and Gemünden 2003; Stremersch et al. 2001; Lusch et al. 2010; Andersen and Drejer 2009). One of the definitions of a service system, proposed by Spohrer et al. (2007), suggests that *value-coproduction configuration of people, technology, other internal and external service systems, and shared information* is the service system itself.

In supply chains, greater control power and value added are typically located closer to the end customer (downstream). (1) The upstream side of the value chain includes raw material suppliers as well as component vendors. (2) Product manufacturers receive the materials by sourcing activities and purchasing. Typically, manufacturers focus on the design and manufacturing of a certain product range. (3) The next level is system integrators, which combine products into systems that will be installed and commissioned for customers. (4) Operations and (5) maintenance services are then the next layers. Communication with the end customer is more frequent, and participation in the value creation process is continuous (Fig. 10.2).

In addition to using servitization as a method to change position in the value chain, technology offers a way of getting closer to operational use of the product and ultimately the end user. Remote technologies are platforms for novel service products. The important value chain-related questions rising from the use of technology in servitization are:

1. Who controls the value chain? When control is given increasingly to a service provider, there is a risk that control of the value chain is affected. A vendor-lock-in situation is not a sustainable solution. The risk of being locked into long-term technological and financial traps is often very real. Another aspect is the transaction cost from one system to another or whether it is possible at all due to physical product ownership. Libraries used to own collections, but now their existence depends on the continuity of journal subscriptions.

10.4 Moving Downstream in a Value Chain

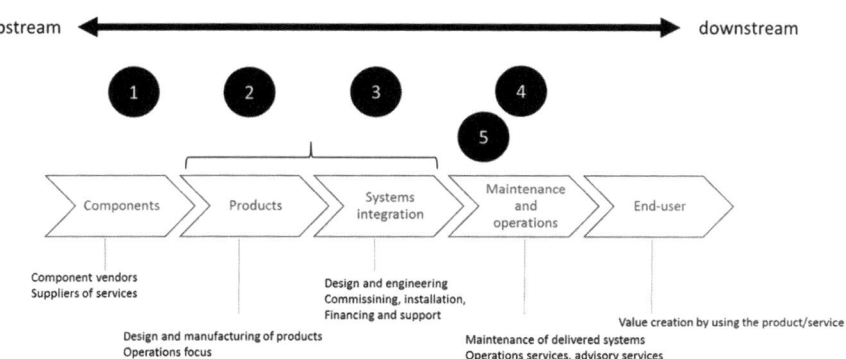

Fig. 10.2 Companies aiming to move in the value chain

2. Who owns the data? A connected system generates a lot of data every day. Data originates from the operational use of the machinery, the ambient environment, as well as the maintenance operations. Production data may be very valuable as it provides information on what has happened at each stage. This information could be very important for competitors or for other parties interested in the financial condition of the operator. Operators of machinery may have also certain rights over the driving data they have generated. Privacy issues may arise.

Tesla Car Supercharger and Battery Swap: From Ownership to Membership
Tesla aims at transforming automotive service. The company started with revolutionizing the idea of a car and followed up with innovative service. Dedicated service centers were located where the highest density of Tesla cars exists. Apart from that, Tesla offers the possibility of an experienced technician performing service work at the customer's premises.

Tesla also offers an innovative approach to charging car batteries by strategically locating charging stations throughout countries. Charging fees are included in the car purchase price and additional charging subscription options can be added. As an alternative Tesla also offers a battery swap, which enables customers to exchange their battery for a fully loaded one in less time that would be needed to fill a fuel tank. Ultimately, customers choose between what is free (supercharger) and what is faster (battery swap) (Tesla Website 2014).

Chapter 11
Conclusions

From the analysis of the literature and industrial cases, it can be concluded that industrial service supply chain management is greatly affected by external and internal factors such as suppliers, employees, customer service management, customer relationship management, and customer feedback. Therefore, it is important to acknowledge that in the case of industrial services, it is not just a product that is being sold. It is also a reputation that plays an important role, and good reputation is built by providing proper services at the right time to the right people.

The focus here should not only be on service innovation and operations mechanism but also on finding solutions to customer's more complex problems beyond the operational issues. Another challenge is improving the implementation of knowledge-based industrial service supply chain management. This can be achieved by maintaining customer relationships and flexibility, which can lead to step-by-step construction and optimization of the service business process system.

The framework proposed by this report aims to highlight the connecting links between the elements. Marketing and operations strategies outline the product, offering, and operational high-level approaches, which are implemented by the product-service systems (PSS). Technology supports PSS and service delivery processes. Pricing decisions are depending from PSS structure. Technology, service delivery, and pricing have an impact on the market and enable desired value chain effects (Fig. 11.1).

The managerial implications of this study encompass the challenges faced by both services and supply chain, concerning how to use scarce resources to match a specific service strategy. Since service strategies are based on the utilization of different types of resources, success will be dependent on the balance between strategy and organizational arrangements. For automotive companies in particular, the choice of service strategy seems to be dependent on the needs of their operations. If the service strategy components from the supplier are critical to production, they would like the supplier to enable the production process through information technology, i.e., pursue a customer service strategy. In contrast, if the complexity or the cost of

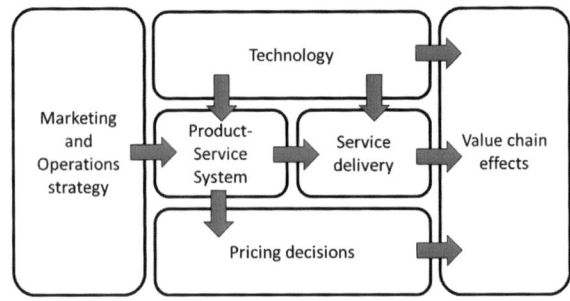

Fig. 11.1 Transition framework for industrial services

the component is high, then the company would like the supplier to participate in the development process, i.e., become a development partner.

Increasing use of software, servitization, and subscription-based pay models are changing how traditional machine-building companies operate. Products are smart and connected. Customers are demanding to see complete solutions instead of partial implementations. ICT combines local presence with remote connections; therefore business planning needs to take the advantage of the possibilities. The examples from industry have already shown that a transformation from physical products to services is possible. The managerial implications are as follows:

1. There is a need for increased focus on customer operations and performance.
2. Key performance metrics should support the service supply chain management.
3. Analysis of operational and maintenance data provides a possibility to find deep understanding of asset utilization and develop new customer solutions.
4. Value-driven pricing—service pricing should produce benefits for the end user.

In general, the focus will gradually shift toward long-term operational expenses of asset utilization. Service products and methods of delivering the service should support this aim.

Acknowledgment

The authors would like to express gratitude to Tekes—the Finnish Funding Agency for Technology and Innovation—for funding OSCMI (Organization and Management of Supply Chains in Industrial Services). The authors would like to thank Dr. Harri Kulmala of FIMECC—the Finnish Metal and Engineering Competence Cluster—for being the chairman of the OSCMI project.

References

11 Crucial Consumer Trends for 2011, Trendwatching Website (2014) http://trendwatching.com/trends/11trends2011/

Andersen PH, Drejer I (2009) Together we share? Competitive and collaborative supplier interests in product development. Technovation 29(10):690–703

Aurich JC, Wolf N, Siener M, Schweitzer E (2009) Configuration of product-service systems. J Manuf Technol Manag 20(5):591–605

Avery SL, Swafford PM (2011) Social capital impact on service supply chains. J Serv Sci 2(2):9–16

Baines TS (2013) Servitization impact study: how UK based manufacturing organisations are transforming themselves to compete through advanced services. Aston Centre for Servitization Research and Practice. Aston Business School, Birmingham

Baines TS, Lightfoot HW, Benedettini O, Kay JM (2009) The servitization of manufacturing: a review of literature and reflection on future challenges. J Manuf Technol Manag 20(5):547–567

Baines T, Lightfoot H, Smart P (2011) Servitization within manufacturing: exploring the provision of advanced services and their impact on vertical integration. J Manuf Technol Manag 22(7):947–954

Barney J (1991) Firm resources and sustained competitive advantage. J Manage 17(1):99–120

Beaumont N (2006) Service level agreements: an essential aspect of outsourcing. Serv Ind J 26(4):381–95

Bitner MJ, Faranda WT, Hubbert AR, Zeithaml VA (1997) Customer contributions and roles in service delivery. Int J Serv Ind Manag 8(3):193–205

Bitner MJ, Ostrom AL, Morgan FN (2008) Service blueprinting: a practical technique for service innovation. Calif Manag Rev 50(3):66

Boonitt S, Pongpanarat C (2011) Measuring service supply chain management processes: the application of the Q-sort technique. Int J Innov Manag Technol 2(3):217–221

Bosch J (2009) From software product lines to software ecosystems. In: Proceedings of the 13th international software product line conference, Carnegie Mellon University, pp 111–119

Boyt T, Harvey M (1997) Classification of industrial services: a model with strategic implications. Ind Mark Manag 26(4):291–300

Burger PC, Cann CW (1995) Post-purchase strategy: a key to successful industrial marketing and customer satisfaction. Ind Mark Manag 24(2):91–98

Buxel H, Esenduran G, Griffin S (2015) Strategic sustainability: creating business value with life cycle analysis. Bus Horiz 58(1):109–122

Buzacott JA (2000) Service system structure. Int J Prod Econ 68(1):15–27

Caterpillar Website (2014) www.cat.com

Chase RB (1978) Where does the customer fit in a service operation? Harv Bus Rev 56:137–42

Chase RB, Aquilano NJ (1992) A matrix for linking marketing and production variables in service system design. Production and Operations Management

Chase RB, Tansik DA (1983) The customer contact model for organization design. Manage Sci 29(9):1037–50

Chen D (2015) A methodology for developing service in virtual manufacturing environment. Annu Rev Control 39:102–117

Clark T (1993) Survey underscores importance of customer service. Bus Market 78(3):41

Coetzee JL (1999) A holistic approach to the maintenance "problem". J Qual Maint Eng 5(3):276–281

Cohen MA, Cull C, Lee HL, Willen D (2000) Saturn's supply chain innovation: high value in after-sales service. Sloan Manage Rev 41(4):93–101

Collier DA (1994) The service/quality solution: using service management to gain competitive advantage. Irwin Professional Publishing, Illinois

Davies A (2004) Moving base into high-value integrated solutions: a value stream approach. Ind Corp Change 13(5):727–756

Edgett S (1994) The traits of successful new service development. J Serv Market 8(3):40–9

Edvardsson B (1997) Quality in new service development: key concepts and a frame of reference. Int J Prod Econ 52(1):31–46

Edvardsson B, Gustafsson A, Roos I (2005) Service portraits in service research: a critical review. Int J Serv Ind Manag 16(1):107–121

Edvardsson B, Holmlund M, Strandvik T (2008) Initiation of business relationships in service-dominant settings. Ind Mark Manag 37(3):339–350

Ellis BA (2008) Condition based maintenance. The Jethro Project 1–5

Ellram LM (1993) A framework for total cost of ownership. Int J Logist Manag 4(2):49–60

Fang E, Palmatier RW, Steenkamp JBE (2008) Effect of service transition strategies on firm value. J Market 72(5):1–14

Fitzsimmons JA, Fitzsimmons M (2004) Service management, 4th edn. Irwin McGraw-Hill, Boston

Freemium.org. (2015) http://www.freemium.org/. Accessed 8 Dec 2015

Froehle CM, Roth AV (2004) New measurement scales for evaluating perceptions of the technology-mediated customer service experience. J Oper Manag 22(1):1–21

Gebauer H (2008) Identifying service strategies in product manufacturing companies by exploring environment–strategy configurations. Ind Mark Manag 37(3):278–291

Gebauer H, Friedli T (2005) Behavioral implications of the transition process from products to services. J Bus Ind Market 20(2):70–8

Gebauer H, Fleisch E, Friedli T (2005) Overcoming the service paradox in manufacturing companies. Eur Manag J 23(1):14–26

Grönroos C (2008) Service logic revisited: who creates value? And who co-creates? Eur Bus Rev 20(4):298–314

Gummesson E (1995) Relationship marketing: its role in the service economy. Underst Serv Manag 244:68

Harland CM (1996) Supply chain management: relationships, chains and networks. Br J Manag 7(s1):S63–S80

Hayes RH, Wheelwright SC (1979) Link manufacturing process and product life cycles. Harv Bus Rev 57(1):133–40

He T, Ho W, Xu XF (2010) A value-oriented model for managing service supply chains. In: 2010 IEEE international conference on Industrial Engineering and Engineering Management (IEEM), IEEE, pp 193–197

Herbert Heinzel, Amit Garg, Stefan Schleyer (2007) Industrial services into manufacturing supply chains

Homburg C, Fassnacht M, Guenther C (2003) The role of soft factors in implementing a service-oriented strategy in industrial marketing companies. J Bus Bus Market 10(2):23–51

References

Hsieh Y, Yuan S (2015) An innovative approach to measuring technology spillovers in service-dominant logic. Kybernetes 44(2):202–219

Hunt SD, Morgan RM (1995) The comparative advantage theory of competition. J Market 59:1–15

Hunt SD (1997) Competing through relationships: grounding relationship marketing in resource-advantage theory. J Market Manag 13(5):431–445

Condition Based Maintenance in Wärtsilä (2014) http://www.wartsila.com/cs/static/flash/studio/assets/content/ss1/cbm-presentation.pdf

Ivens BS (2005) Flexibility in industrial service relationships: the construct, antecedents, and performance outcomes. Ind Mark Manag 34(6):566–576

Jo Bitner M, Faranda WT, Hubbert AR, Zeithaml VA (1997) Customer contributions and roles in service delivery. Int J Serv Ind Manage 8(3):193–205

Johansson P, Olhager J (2004) Industrial service profiling: matching service offerings and processes. Int J Prod Econ 89(3):309–20

Joy Global Website (2014) www.joy.com

Karwan KR, Markland RE (2006) Integrating service design principles and information technology to improve delivery and productivity in public sector operations: the case of the South Carolina DMV. J Oper Manage 24(4):347–62

Kellogg DL, Nie W (1995) A framework for strategic service management. J Oper Manage 13(4):323–37

Kindström D, Kowalkowski C (2009) Development of industrial service offerings: a process framework. J Serv Manag 20(2):156–172

Kindström D, Kowalkowski C (2014) Service innovation in product-centric firms: a multidimensional business model perspective. J Bus Ind Mark 29(2):96–111

Kohtamäki M, Partanen J, Möller K (2013) Making a profit with R&D services—the critical role of relational capital. Ind Mark Manag 42(1):71–81

Kottler JA (1994) Advanced group leadership. Brooks/Cole Publishing Company, Pacific Grove

Kowalkowski C (2010) What does a service-dominant logic really mean for manufacturing firms? CIRP J Manuf Sci Technol 3(4):285–292

Kowalkowski C, Kindström D, Brehmer PO (2011) Managing industrial service offerings in global business markets. J Bus Ind Mark 26(3):181–192

Krikke HR, van Nunen JAEE, Zuidwijk RA, Kuik R (2004) E-business and circular supply chains: increased business opportunities by IT-based Customer oriented Return-flow Management. In: Fleischmann B, Klose A (eds) Distribution logistics: advanced solutions to practical problems, vol 544, Lecture notes in economics and mathematical systems. Springer, Berlin, pp 221–242

Lin Y, Shi Y, Zhou L (2010) Service supply chain: nature, evolution, and operational implications. In: Proceedings of the 6th CIRP-sponsored international conference on digital enterprise technology, Springer, Berlin, pp 1189–1204

Lindahl M, Sundin E, Sakao T (2014) Environmental and economic benefits of integrated product service offerings quantified with real business cases. J Clean Prod 64:288–296

Löfberg N, Witell L, Gustafsson A (2010) Service strategies in a supply chain. J Serv Manag 21(4):427–440

Logicalis Australia (2015) Understanding CAPEX vs OPEX for technology spending. https://logicalisaus.wordpress.com/2013/12/09/understanding-capex-vs-opex/. Accessed 8 Dec 2005

Lovelock CH (1983) Classifying services to gain strategic marketing insights. J Market 47(3):9–20

Lusch RF, Vargo SL, O'Brien M (2007) Competing through service: insights from service-dominant logic. J Retail 83(1):5–18

Lusch RF, Vargo SL, Tanniru M (2010) Service, value networks and learning. J Acad Mark Sci 38(1):19–31

Maintenance Assistant Website (2014) http://www.maintenanceassistant.com/condition-based-maintenance/

Marilly E, Martinot O, Betgé-Brezetz S, Delègue G (2002) Requirements for service level agreement management. In: IP operations and management, 2002 IEEE workshop on, IEEE, pp 57–62

Mathe H, Shapiro RD (1993) Integrating service strategy in the manufacturing company. Chapman & Hall, London

Matthyssens P, Vandenbempt K (2008) Moving from basic offerings to value-added solutions: strategies, barriers and alignment. Ind Mark Manag 37(3):316–328

Mc Arthur R (n.d.) The real meaning of enterprise asset maintenance. http://www.genesissolutions.com/wp-content/uploads/2009/10/GenesisSolutions-The-Real-Meaning-of-EAM.pdf. Accessed 8 Dec 2015

McLaughlin CP, Fitzsimmons JA (1996) Strategies for globalizing service operations. Int J Serv Ind Manag 7(4):43–57

Metters R, Vargas V (2000) A typology of de-coupling strategies in mixed services. J Oper Manage 18(6):663–82

Mont OK (2002) Clarifying the concept of product–service system. J Clean Prod 10(3):237–245

Morgan RM, Hunt SD (1994) The commitment-trust theory of relationship marketing. J Market 58:20–38

Morris MH, Davis DL (1992) Measuring and managing customer service in industrial firms. Ind Mark Manag 21(4):343–353

Neely A, Benedettini O, Visnjic I (2011) The servitization of manufacturing: further evidence. In: 18th European Operations Management Association conference, Cambridge, pp 3–6

Oliva R, Kallenberg R (2003) Managing the transition from products to services. Int J Serv Ind Manag 14(2):160–172

OPEX and CAPEX Comparison, Diffen Website (2014) http://www.diffen.com/difference/Capex_vs_Opex

Osterwalder A, Pigneur Y (2010) Business model generation: a handbook for visionaries, game changers, and challengers. Wiley, Hoboken

Paiola M, Saccani N, Perona M, Gebauer H (2012) Moving from products to solutions: strategic approaches for developing capabilities. Eur Manag J 2(013):31

Parasuraman A, Berry LL, Zeithaml V (1991) Understanding, measuring and improving service quality: Findings from a multiphase research program. In: Brown W, Gummesson E, Edvardsson B, Gustavsson B (eds) Service quality. multidisciplinary and multinational perspectives. Lexington Books, Lexington, pp 253–268

Partanen J, Kohtamäki M (2011) SERVSCOPE—a measurement method for the scope of industrial services. In: EURAM2011—business conference

Paschke A, Schnappinger-Gerull E (2006) A categorization scheme for SLA metrics. Serv Orient Electron Commer 80:25–40

Payne A, Storbacka K, Frow P, Knox S (2009) Co-creating brands: diagnosing and designing the relationship experience. J Bus Res 62(3):379–389

Ponsignon F, Smart PA, Maull RS (2011) Service delivery system design: characteristics and contingencies. Int J Oper Prod Manag 31(3):324–349

Porter M (2001) The value chain and competitive advantage. In: Barnes D (ed) Understanding business: processes. Psychology Press, New York

Prasad BVS, Selven K (2010) Supply chain management in service industry. ICFAI University Press, Hyderabad

Rabetino R, Kohtamäki M, Lehtonen H, Kostama H (2015) Developing the concept of life-cycle service offering. Ind Mark Manag 49:53–66

Raddats C, Burton J (2011) Strategy and structure configurations for services within product-centric businesses. J Serv Manag 22(4):522–539

Raddats C, Kowalkowski C (2014) A reconceptualization of manufacturers' service strategies. J Bus-To-Bus Mark 21(1):19–34

Raja J, Bourne D, Goffin K, Çakkol M, Martinez V (2013) Achieving customer satisfaction through integrated products and services: an exploratory study. J Prod Innov Manag 30(6):1128–1144

Rehme J (n.d.) Prices and contract for service selling. http://www.mtcstiftelsen.se/Uploads/Files/22.pdf

Reid RD, Sanders NR (2005) Operations management: an integrated approach. Wiley, Hoboken

References

Reim W, Parida V, Örtqvist D (2015) Product–service systems (PSS) business models and tactics—a systematic literature review. J Clean Prod 97:61–75
Ren G, Gregory M (2007) Servitization in manufacturing companies. In: 16th frontiers in service conference. San Francisco, CA
Ritter T, Gemünden HG (2003) Network competence: its impact on innovation success and its antecedents. J Bus Res 56(9):745–755
Robotis A, Bhattacharya S, Van Wassenhove LN (2012) Lifecycle pricing for installed base management with constrained capacity and remanufacturing. Prod Oper Manag 21(2):236–252
Rolls-Royce Website (2014). http://www.rolls-royce.com/media/press-releases/yr-2012/121030-the-hour.aspx
Roth AV, Menor LJ (2003) Insights into service operations management: a research agenda. Prod Oper Manage 12(2):145–64
Samli AC, Jacobs LW, Wills J (1992) What presale and postsale services do you need to be competitive. Ind Market Manage 21(1):33–41
Samli AC, Wills JR, Jacobs L (1993) Developing global products and marketing strategies: a rejoinder. J Acad Market Sci 21(1):79–83
Sasser EW, Olsen PR, Wyckoff DD (1978) Management of service operations: text, cases, and readings. Allyn & Bacon, Boston
Schmenner RW (2009) Manufacturing, service, and their integration: some history and theory. Int J Oper Prod Manag 29(5):431–443
Schmitt BH (2003) Customer experience management: a revolutionary approach to connecting with your customers. Wiley, Hoboken
Sharma A, Lambert DM (1994) How accurate are salespersons perceptions' of their customers? Ind Market Manage 23(4):357–65
Shostack GL (1987) Service positioning through structural change. J Market 34–43
Shohet IM (2003) Building evaluation methodology for setting maintenance priorities in hospital buildings. Constr Manag Econ 21(7):681–692
Silvestro R (1999) Positioning services along the volume—variety diagonal. Int J Oper Prod Manage 19(3/4):399–420
Silvestro R, Fitzgerald L, Johnston R, Voss C (1992) Towards a classification of service processes. Int J Serv Ind Manage 3(3):62–75
Spohrer J, Maglio P, Bailey J, Gruhl D (2007) Steps toward a science of service systems. Computer 40(3):71–7
Stanton W, Etzel M, Walker B (1991) Fundamentals of marketing, 9th edn. McGraw-Hill, Inc., New York
Storbacka K, Windahl C, Nenonen S, Salonen A (2013) Solution business models: transformation along four continua. Ind Mark Manag 42(5):705–716
Strandvik T, Holmlund M, Edvardsson B (2012) Customer needing: a challenge for the seller offering. J Bus Ind Market 27(2):132–41
Strassner EH, Howells TF III (2005) Annual industry accounts: advance estimates for 2004. Surv Curr Bus 85:7–19
Stremersch S, Wuyts S, Frambach RT (2001) The purchasing of full-service contracts: an exploratory study within the industrial maintenance market. Ind Mark Manag 30(1):1–12
Subscription Economy, Fortune Website (2014) http://fortune.com/2014/06/06/welcome-to-the-subscription-economy/
Subscription Economy, Inc. Website (2014) http://www.inc.com/bo-burlingham/why-john-warrillow-is-all-about-subscription-services.html
Techtarget Website (2014) http://searchitchannel.techtarget.com/feature/What-are-the-popular-pricing-models-for-managed-services-providers
Tesla Website (2014) http://www.teslamotors.com/
Tukker A (2013) Product services for a resource-efficient and circular economy—a review. J Clean Prod 79:76–91
Tuli KR, Kohli AK, Bharadwaj SG (2007) Rethinking customer solutions: from product bundles to relational processes. J Mark 71(3):1–17

Van Looy B, Gemmel P, Dierdonck R (2003) Services management: an integrated approach. Pearson Education, Harlow

Vandermerwe S, Rada J (1989) Servitization of business: adding value by adding services. Eur Manag J 6(4):314–324

Vargo SL, Lusch RF (2008) From goods to service(s): divergences and convergences of logics. Ind Mark Manag 37(3):254–259

Vargo SL, Maglio PP, Akaka MA (2008) On value and value co-creation: a service systems and service logic perspective. Eur Manag J 26(3):145–152

Vukelja E, Runje B (2014) Quality service evaluation through the system of complaints and praise. Interdiscip Descr Complex Syst 12(1):78–91, Academic Search Elite, EBSCO*host*. Accessed 9 Feb 2016

Wemmerlöv U (1990) A taxonomy for service processes and its implications for system design. Int J Serv Ind Manage 1(3):20–40

Wemmerloev U (1990) A taxonomy for service processes and its implications for system design. Int J Serv Ind Manage 1(3):20–40

West S, Pascual A (2015) The use of equipment life-cycle analysis to identify new service opportunities. In: Proceedings of the spring servitization conference (SSC2015)

Wouters JP (2004) Customer service strategy options: a multiple case study in a B2B setting. Ind Market Manage 33(7):583–92

Zeithaml VA, Bitner MJ, Gremler DD (2006) Services marketing: integrating customer focus across the firm. McGraw-Hill, Irwin

Index

A
Assets, industrial services, 27

B
Big data, 69–70

C
CAPEX. *See* Capital expenditures (CAPEX)
Capital expenditures (CAPEX), 75–77
CBM. *See* Condition-based maintenance (CBM)
Condition-based maintenance (CBM)
   assets, 65
   benefits, 65
   disadvantages, 65, 66
   O&M and ISO 13774, 65
   objective, 65
   performance indicators, 65
   phases, 66, 67
   types, 66, 67
   Wärtsilä, 66, 68
Cost
   customer acquisition, 54
   delivery, 51
   of failure, 55
   installations, 53
   savings, 54

E
EAM. *See* Enterprise asset management (EAM)
Enterprise asset management (EAM), 55–56

F
Freemium, 79, 80

G
Goods-dominant (G-D) to service-dominant (S-D) logic
   comparison, 34–35
   guidelines, 33, 34
   transition, 33, 34

H
Horizontal integration, 84–86

I
Industrial services
   business model, 33
   classification scheme, 29, 30
   competition, 27
   dimensions, service systems, 29
   disturbances, 28
   efficiency, 27
   factors, 89
   features, 28
   flexibility, 27
   generic and wide, 28
   goods-dominant to service-dominant logic, 34–35
   methods, classification, 28, 29
   products, 27, 31, 32, 90
   properties, 28
   resources, 27

Industrial services (*cont.*)
  service capacity and performance, 28
  service delivery types, 30, 31
  service strategies, 89
  servitization strategies, 31
  SERVSCOPE, 30
  solutions, 89
  transition framework, 89, 90
  variability, 28
Installed base management
  activities, 53
  benefits, 54
  definition, 53, 54
  EAM, 55–56
  marketing and operations, 54
  pricing contracts, 53
  product-oriented services, 55
  profit, 53

M
Managed service providers (MSPs), 78
Marketing and operation strategy
  competitiveness, service-dominant logic, 38, 39
  co-production of services, 37, 38
  customer experience management, 37
  global service strategies, 38–40
  service supply chain structure, 41–42
Mobile applications
  characteristics, 71
  features, 70
  machine vendors, 71
MSPs. *See* Managed service providers (MSPs)

N
New service development (NSD), 11–15

O
Operating expenditure (OPEX), 75–77
OPEX. *See* Operating expenditure (OPEX)

P
Performance measurement, 51–53
Pricing models
  capability, 77
  freemium, 79, 80
  Joy Global, 78, 79
  MSPs, 78
  services, 77, 78

Product life cycle
  machinery delivery, 15–16
  maintenance services and extended warranties, 15
  software process life cycle, 16–18
Product-service systems (PSS)
  barriers, 23–24
  benefits, 21–22
  characteristics, 22–23
  configuration, 20, 21
  customer assets, 1
  definition, 9
  industrial customer services, 2
  issues and complexities, 2
  relationship marketing, 2
  remote monitoring systems, 1
  revenue shares, global industries, 3
  service component, 20
  service economy, 1
  service supply chain, 1
  servitization, 2
  software and information component, 20
  stages, service provider, 24
  supply chain management, 3
  transition, 4
  types, industrial services, 19, 20
PSS. *See* Product-service systems (PSS)

R
Remote management
  applications, 60
  centralized servers, 61
  operations and maintenance (O&M), 61
  product life cycle, 60
  site support functionalities, 61
  user interfaces, 61, 62
Risk, 23, 24, 38, 45, 58, 78, 86

S
Service delivery, 49–51
  concept, 43, 44
  customer expectation, 45–47
  customers' roles, 44–46
  gap-analysis, service quality, 52, 53
  installed base management, 53–55
  interactions, 43
  methods, 45
  performance measurement, 51–53
  SLAs (*see* Service-level agreements (SLAs))
  system design, 43–44, 46

Service design, 11, 43
Service-level agreements (SLAs), 19, 20
  definition, 49, 50
  escape clauses/constraints, 50
  life cycle, 50, 51
  monitoring and reporting, 50
  pricing, 50
  reliability and responsiveness, 50
  service provider, 50
  service quality, 50
  service schedule, 50
Service supply chain (SSC), 1, 37, 41–42, 89, 90
Service quality, 47, 52–54
Servitization
  commodization, 9, 10
  decommodization, 10
  definition, 5, 6, 8–9
  features, 5, 9
  goods and services, products, 6
  interactions, 5
  manufacturing and service organizations, 10–11
  manufacturing companies, 7, 8
  NSD, 11–15
  paradox, 18
  product life cycle, 15–18
  products and services, characteristics, 6, 7
  service blueprinting, 12–13
  service components, 11
  service innovation and business models, 13–15
  service types, 6
  spectrum, 6
  transformation, 7, 8
SLAs. *See* Service-level agreements (SLAs)
Software
  analytical and controlling functions, 59
  ecosystem, 58
  updates, 69
Software ecosystems
  directed approach, 80
  tiers of developers, 81
  transition, 79
  undirected approach, 81
Subscription
  economy, 77
  ownership
    consumer services, 73
    forms, 74, 75
    innovative approaches, 73, 74
    payment models, 73
    spectrum, 74
Supplier, 68

T
Technology and servitization
  CBM, 64–68
  cloud-based services and portals, 68–69
  customer contact, 57, 58
  industrial standards, service platforms
    bottom-up process, CBM, 63
    business opportunities, 63
    development, 63
    service products, remote management, 63, 64
  internet and connected products, 60–62
    architecture, smart physical products, 59–60
    potential implication strategies, 58
    remote management (*see* Remote management)

U
Utilization
  customer asset, 52
  value cocreation, 84

V
Value chain effects, 4
Value chains
  cocreation and coproduction, 83–84
  definition, 83
  downstream, 86–87
  supply chains and networks, 84, 85
  value proposition, 83
  vertical and horizontal integration, 84–86
Value offering, 84
Value proposition, 83
Vendors
  product, 2
  service, 4
Vertical integration, 84–86

W
Wärtsilä CBM service, 66, 68

Printed in Great Britain
by Amazon